Psychological Operations American Style

Psychological Operations American Style

The Joint United States Public Affairs Office, Vietnam and Beyond

Robert J. Kodosky

LEXINGTON BOOKS

A division of
ROWMAN & LITTLEFIELD PUBLISHERS, INC.
Lanham • Boulder • New York • Toronto • Plymouth, UK

LEXINGTON BOOKS

A division of Rowman & Littlefield Publishers, Inc.
A wholly owned subsidiary of The Rowman & Littlefield Publishing Group, Inc.
4501 Forbes Boulevard, Suite 200
Lanham, MD 20706

Estover Road
Plymouth PL6 7PY
United Kingdom

British Library Cataloguing in Publication Information Available

Library of Congress Cataloging-in-Publication Data

Psychological operations American style : the Joint United States
Public Affairs Office, Vietnam and beyond / Robert J. Kodosky.
p. cm.
ISBN-13: 978-0-7391-2139-9 (cloth : alk. paper)
ISBN-10: 0-7391-2139-1 (cloth : alk. paper)
1. United States. Army. 2. Psychological warfare—United States—History.
UB276 .K64 2007
355.3/4340973–dc22 2007028974

Printed in the United States of America

∞™ The paper used in this publication meets the minimum requirements of American National Standard for Information Sciences—Permanence of Paper for Printed Library Materials, ANSI/NISO Z39.48-1992.

For Linda Lee

Contents

Preface

When people learn about what I do, the first question is always "Why Vietnam?" Of course, they have no idea that very question served as the title of a booklet published by the USIA to answer that question on behalf of the United States government back in the 1960s. Regardless, my answer is that my interest in the war derives from growing up in its aftermath. As a child in Portage, Pennsylvania, during the 1970s, one did not often get to hear much from adults about such a divisive conflict that impacted the lives of so many in our small town. In school, whether by design or not, we ended every year with the Second World War. Consequently, when I arrived at college in the 1980s, I set out to investigate this recent event that now looked as distant as the stone age yet continued to haunt the Americans who lived through it, their families and friends, and the nation as a whole. Sadly, as I've learned through the course that I teach on the war, to history majors no less, the war in Vietnam remains as inaccessible to young people as ever. While many of them have viewed movies such as *We Were Soldiers*, few of my students, at least have had much exposure to the war in the classroom prior to our meetings.

When people ask about PSYOP, which happens far more rarely, my answer is similar to the one I give regarding Vietnam. Particularly as a graduate student, PSYOP and public diplomacy always appeared lurking at the margins of traditional military and diplomatic history. These practices provided historians with great anecdotes, but, until recently attracted little scholarly attention. The end of the Cold War and the subsequent onset of the War on Terrorism have resulted in a number of fine studies considering the relationship between America's foreign policy and its efforts to conduct PSYOP and public diplomacy. I have benefited from these works immensely in pursuing my own study of the Joint United States Public Affairs Office (JUSPAO) in Vietnam.

Of course, I have also had the great fortune of studying under, and working with, some tremendous scholars who have provided me with the proper mix of encouragement and criticism. At Temple University, my debt is especially immense. As a graduate student there, I worked in an amazingly constructive environment and benefited from the feedback of the many wonderful students and faculty that I encountered over the years. I am especially grateful for all of the help so freely given to me by Jay Lockenour, Dieu Nguyen, and my dissertation supervisor, eminent diplomatic historian Richard H. Immerman. These three esteemed scholars and teachers gave me more than I could ever hope to repay. They read numerous drafts of this manuscript and shared with me their considerable insights. They somehow sensed whatever I most needed at any given moment, a helping hand or a kick in the pants, and provided it to me. I am also grateful to Nancy Snow, a former USIA official and current communications scholar, for lending me her considerable expertise by serving as an outside reader. In addition to the generous contributions of these individuals, I am most thankful for the guidance I received during much of my time at Temple from the late military historian Russell F. Weigley, truly the model of a great teacher and scholar. He is sorely missed.

At West Chester University where I teach, I have received enormous amounts of support and feedback from my colleagues, particularly William Hewitt, Claude Foster, and Jim Jones. My current chair, Thomas Legg, and his predecessor Richard Webster, have consistently expressed faith in my ability to work as a historian and teacher. Of course, where would any of us there be without our department's secretary Carolee Marciano. I am also most grateful for the students who have enrolled in my course at West Chester on the Vietnam War. They have been a source of great inspiration for my work, invigorating me with their passion for the subject. Along the way, I have had a great deal of help from others, including, but surely not limited to, Marc Gallicchio at Villanova University, John Baky at LaSalle University, Marc Jason Gilbert at North Georgia College, Robert Stephan at the Institute of World Politics, Robert Steele at OSS.Net, and Steve Maxner at Texas Tech University, and of course, all of the individuals at Lexington Books who have helped me bring this project to its completion. In addition, the staff at the various archives that I have visited; the Lyndon Baines Johnson Presidential Library at the University of Texas in Austin, Texas, the Military History Institute at Carlisle, Pennsylvania, and the Vietnam Center at Texas Tech University in Lubbock, Texas, have been nothing short of extraordinary.

My work has been aided by many individuals along the way, but all of the omissions and mistakes it contains are wholly my own. Those people who are most likely to forgive all of my oversights are the family and friends who have patiently waited for me to emerge from my office above the garage time and time again to get on with the truly important things in life. Unfortunately, three of my most ardent supporters, my mom Jackie, my father-in-law Joe, and his mother Teresa, are no longer with us. They are missed and remembered daily.

At home, my canine companions, Max, Lady, and Tess have never ceased to remind me of the rejuvenating potential contained in mid-afternoon field exercises. Every author should be so lucky to have such a platoon always at the ready residing at chair's edge. Finally and most importantly, this study simply would not exist without the support and encouragement provided to me always by my wife Linda. She has been there every step of the way, fully believing me capable of taking the next one when often I thought it not possible. It is to her that I dedicate this book.

Introduction

SHOCK AND AWE

On a crisp autumn morning in 2001, I walked into a newsstand just outside of Philadelphia, Pennsylvania, to fulfill my daily need for current events and caffeine. As I approached the counter, alone in the store, I noticed that the proprietor's eyes remained transfixed on the television set behind me, which provided the latest updates from the Cable News Network (CNN) on Operation Enduring Freedom—Afghanistan (OEF-A). Without even looking at me he asked, "You know what we ought to do?" Before I could reply, he offered, "Nuke 'em." Perhaps sensing my disappointment in his proposed action, the store owner now fixed his eyes on me and said, "Before that, though, we ought to send a few planes full of pigs over there." He continued, "Damn Towel-Heads are terrified of pigs, they will come running out of those tunnels for sure."

"Shock and awe" is a phrase that well describes my reaction. In hindsight, I probably should have offered this man a thoughtful response to his planned pig-army, one that contained the sage advice that then United States Army Lieutenant General David H. Petraeus provided in the pages of *Military Review*. Based on his experience in Iraq, now Major General Petraeus counsels, "Cultural awareness is a force multiplier . . . people are, in many respects, the decisive terrain, and we must study that terrain in the same way that we have always studied the geographic terrain."[1] An enlightened perspective to be sure, but it is not one that I managed to share with my friend in the newsstand that day. Instead, I could only leave the store in silence, lost in thought considering precisely how close this man had come to articulating the traditional style of American psychological operations (PSYOP).[2]

This mentality has spurred planners to unleash foxes in Japan, vampires in the Philippines, forest demons in Vietnam and supernatural tanks in the Middle

East. It has produced PSYOP more rooted in fertile imaginations than in rich intelligence. It has served America well in support of conventional military tactics, but it has often generated appeals to win hearts and minds that have gone lost in translation with their intended audiences.

As currently defined by the United States Department of Defense (DOD), psychological operations (PSYOP) consist of "planned operations to convey selected truthful information and indicators to foreign audiences to influence their emotions, motives, objective reasoning, and ultimately, the behavior of their governments, organizations, groups, and individuals." Their purpose is "to induce or reinforce foreign attitudes and behavior favorable to the originator's objectives." DOD considers PSYOP "a vital part of the broad range of US activities to influence foreign audiences." It possesses exclusive authorization "to influence foreign [target audiences] directly through the use of radio, print, and other media."

Along with electronic warfare (EW), computer network operations (CNO), military deception (MILDEC), and operations security (OPSEC), PSYOP is a core element of United States Information Operations (IO). Along with Public Affairs (PA), International Broadcasting (IB), and Public Diplomacy (PD), IO comprises the nation's Strategic Communications process. The aim of this apparatus is to "*understand* global attitudes and cultures, *engage* in a dialogue of ideas between people and institutions, *advise* policy makers, diplomats, and military leaders on the public opinion implications of policy choices, and *influence* attitudes and behavior through communications studies."[3]

The United States specifically utilizes IO "to influence, disrupt, corrupt, or usurp adversarial human and automated decision making while protecting our own."[4] This use reflects the American military's increasing awareness that in modern warfare, as British Broadcasting Company (BBC) news director Kenneth Payne observes, winning is "as much dependent on carrying domestic and international opinion as it is on defeating the enemy on the battlefield."[5]

When compared to the present definition, the way in which Americans conceptualized PSYOP in Vietnam is strikingly simple. Officially, they described it as "the planned use of propaganda and other measures to influence the opinions, emotions, attitudes, and behavior of hostile, neutral, or friendly groups in such a way as to support the achievement of national objectives."[6] As construed by the Joint United States Public Affairs Office (JUSPAO), the civilian-led agency created in 1965 to guide American PSYOP in Vietnam, its mission consisted of "nothing more—and nothing less—than the whole process of communications in wartime."[7] Under JUSPAO guidance, the United States Army's 4th PSYOP Group (formerly 6th PSYOP Battalion), the "workhorse of American PSYOP in Vietnam," carried out myriad campaigns designed to win Vietnamese support for the Republic of Vietnam (RVN), the anticommunist government in Saigon that Americans endorsed as an alternative to the Democratic Republic of Vietnam (DRV) in Hanoi.

As President Lyndon Baines Johnson often expressed regarding Vietnam, American PSYOP is a recognition that ". . . the ultimate victory will depend upon the hearts and the minds of the people who live out there."[8] To understand how the United States has acted on that awareness necessitates situating its PSYOP within the American Way of War identified by the late Russell F. Weigley, an eminent United States military historian. In his seminal 1973 book, Weigley argued that a distinctive American Way of War evolved from the Indian Wars through the Civil War and reached its zenith during the Second World War. The United States mobilized its considerable material and human resources to annihilate the armies of their enemies, to destroy the moral and economic capacities of their opponents to make war, and to secure unconditional surrender at negotiating tables. For much of America's history, this way of warfare served United States interests very well.[9]

PSYOP—AMERICAN STYLE

While a graduate student at Temple University, I explored how PSYOP fit into this American Way of War and offered Professor Weigley my analysis in a seminar paper. A few days after submitting my thesis, to my great surprise as I made my way across campus, I heard Professor Weigley call me out by name. As I humbly stood before him, this great scholar informed me that he loved the paper, especially "the Rodney Dangerfield part." "You are right," he said, "just like the comedian, PSYOP gets no respect."

As I suggested in that paper, winning hearts and minds hardly proved compatible with the explicit goal of the enemy's utter destruction advanced by the American Way of War. Out of necessity, the nation's early leaders, exemplified by General George Washington, demonstrated a mastery of PSYOP. As American military might increased, though, the need for leaders to utilize persuasion considerably lessened. To the extent that the United States adopted PSYOP to complement its preferred means of pursuing combat, it supported the nation's aggressive mentality. Commanders considered PSYOP a tactical weapon that, at best, they could use to instill fear and terror that might hasten the collapse of enemy resistance. At worst, they considered it as a drain on resources that unnecessarily slowed their primary task of destroying their enemy's army. They valued PSYOP only to the extent that it produced measurable effects when applied as a weapon of war.

Vietnam proved an ill fit for the American Way of War and the style of PSYOP that it had spawned. Political and military restraints prevented the

United States from pursuing the unconditional surrender of its adversaries through full scale application of military might in the familiar style. The dissonance at home that the conflict generated sprang from the nation's inability to wage war in the manner its public had come to perceive as the natural approach to the problem of armed conflict. In Professor Weigley's view, Vietnam ushered in a new era of warfare that demanded from America "not overwhelming weight of resources," but "light, agile, maneuverable, politically sensitive and sophisticated armed forces for peacemaking and peace keeping roles."[10]

Implicit in this prescriptive analysis for the United States is the need for a more sophisticated approach to PSYOP, one that is proactive rather than reactive, and one that meets strategic needs as well as tactical ones. In other words, as expressed recently by United States Army Major General Peter W. Chiarelli and Major Patrick R. Michaelis, American PSYOP must contribute to a broader communications strategy that successfully anticipates each area of operation's cultural realities. The idea is for the United States to incorporate fully its IO into a military plan designed not only to win the war but also "to win the peace."[11]

LESSONS LOST

There is little evidence to suggest that the United States fulfilled this notion in Vietnam. In fact, besides the astute observations offered by a few individuals such as Gen. Chiarelli and Maj. Michaelis, it is unclear whether officials have learned from, or if they have even yet sought to assess, the lessons offered by America's use of PSYOP in Vietnam. For example, according to United States Army Colonel Curtis D. Boyd, "PSYOP has devolved into a pejorative term both inside and outside our military." Citing lessons learned from American operations in Afghanistan and in Iraq, along with recent Defense Science Board studies, PSYOP master plans, and a National Defense University study, Boyd concludes that the United States continues to lack "sufficient force, capabilities, and authorities to inform and influence to an adequate degree foreign populations when and where we desire."[12]

Others have offered equally bleak evaluations. Edward P. Djerejlan, Chairman, Advisory Group of Public Diplomacy for the Arab and Muslim World, testified before a congressional committee on February 4, 2004, that "The bottom has indeed fallen out of support for the United States." Djerejlan further cautioned that the problem is "not limited to the Arab world." In Spain, an American ally in Iraq, nearly forty percent of the population held an unfavorable view of American policy.[13] Even one of the chief architects of Operation Iraqi Freedom (OIF), former Secretary of Defense Donald H. Rumsfeld, informed his audience at the United States Army War College in Carlisle, Pennsylvania, on March 23, 2006, "If I were grading, I would say we probably deserve a D or a D-plus as a country as to how well we're doing in the battle of ideas that's taking place in the world today."[14]

The noted political psychologist Jerrold M. Post, founder and former director of the Central Intelligence Agency's Center for the Analysis of Personality and Political Behavior, points out that "PSYOP should be the primary weapon in the war against terrorism." As he concedes, though, "there has been little attention" paid by American officials to "the potential of strategic PSYOP in undermining the enemy to prepare the battlefield."[15]

The old "Rodney Dangerfield" mentality clearly still exists. According to one recent report:

> During the IO Symposium of 15–16 December 2005 at Fort Leavenworth, Kansas, one of the participants remarked candidly, "If you tell a commander he can win without integrating IO, then he won't bother to integrate IO." Given that unit training could not easily support IO integration, commanders and their staffs needed to receive more in depth training on IO as part of their professional development.[16]

The Army remains a full forty percent short of required IO personnel, though, with the expectation of fulfilling all requirements only by 2013. America's PSYOP force has recently undergone a major realignment. Two reserve PSYOP groups formerly under U.S. Army Special Operations Command (USASOC) have been reassigned to Army Reserve Command (USARC). While this move should be a step toward achieving the integration of PSYOP into conventional forces, the Army's only active group that remains under USASOC is the 4th. Moreover, a full two-thirds of Army PSYOP continues to reside within Army Reserve. As United States Army Colonel Curtis D. Boyd suggests, "The realities of the foreign communication challenges we face demand even greater Army structural and organizational change." No less than "'Total Army' participation" is required.[17]

This conditioner is the case especially since shortcomings in PSYOP Intelligence remain glaring. A recent Rand Study on Intelligence support to IO during OIF determined that Intelligence failed in providing field commanders with the level of support they required to conduct their operations. It found that "although intelligence doctrine implied support to IO, there was no systematic method or analytical model for providing intelligence support to IO."[18] According to United States Army Captains Stoney Trent and James L. Doty III:

> A significant shortfall in military operations continues to be insufficient knowledge about the local populace and how to influence it. Combat units and intelligence

> sections in the 1st Armored Division and 3d Infantry Division deployed to Iraq with country studies printed before 1991, for Operation Desert Storm. These studies included only general population information. Well-developed information about tribes, such as their motivation, leaders, interrelationships or even general locations, simply did not exist.[19]

In an incident that occurred on October 1, 2005, American soldiers in Gumbad, Afghanistan, decided to dispose of the bodies of two enemy fighters killed in a firefight the previous day by burning them. Despite Islamic custom that forbids cremation, the American officer in charge made the decision to proceed with the incineration due to what he later identified as "hygienic reasons." When apprised of the act, the unit's battalion commander ordered the flaming bodies extinguished. The troops complied immediately. An official investigation of the affair issued a report in November 2005 that found evidence of "poor decision-making and judgment, poor reporting and a lack of knowledge and respect for local Afghan custom and tradition."

In total, four American soldiers received "administrative punishment" for what the Army identified as two separate incidents. The first case involved the individuals responsible for the act itself. The second, however, implicated two PSYOP specialists who heard about the incident while operating in the area and decided to use it in hopes of inciting Taliban fighters. They broadcast information about the cremation over loudspeakers aimed "toward the village and toward the local mountains in which enemy forces were suspected of hiding." While investigators found "no evidence the bodies were burned for psychological effect," they concluded that the broadcasts violated "standing policies for the content of loudspeaker messages" and urged that all troops in the command undergo training on "Afghan sensitivities."[20]

Over six years into its War on Terrorism, and over thirty years removed from the Vietnam War, the United States continues to cling to its traditional style of PSYOP, a type that took its shape from the historical American Way of War. PSYOP has remained a tangential weapon in the otherwise conventional arsenal employed by United States officials in the War on Terrorism. To the extent Americans have utilized PSYOP they have remained wedded to the notion of its use as a tactical offensive weapon, one which instills terror in their enemies. While often successful in the short term for securing defection and surrender, this type of PSYOP does little to win hearts and minds over the long haul.

As America's experience in Vietnam demonstrates, when conceptualized primarily as a tactical offensive weapon, PSYOP possesses the potential to undermine the nation's position by eroding its credibility. It offers civilian officials and military commanders the means to blur the distinction between information and persuasion in order to achieve immediate and demonstrable results. Officials learned this lesson in Vietnam, but not until after they squandered their credibility by authorizing the JUSPAO both to inform and persuade its audiences of American policies. The agency orchestrated American PSYOP even as it conducted official press conferences for much of the war, leaving it open to being charged by critics as constituting little more than a propaganda department.

Current American officials have come dangerously close to making the same mistake made by their predecessors. In 2002, the Pentagon attempted to establish the Office of Strategic Influence (OSI) on a propagandistic information model. A media backlash quickly thwarted this effort, but only two years later another initiative surreptitiously placed articles in newspapers by the Lincoln Group. Canadian Armed Forces Commander J. D. Scanlon contends that politicians and military commanders who are seeking "more precise effects on the battlefield" are "blurring the boundaries between IO and PA." As a result, they risk aiding and abetting "the enemy's portrayal of America as a hypocritical interloper."[21] Scanlon's analysis is insightful, and it is reaffirmed through the study of the style of PSYOP employed by America in Vietnam.

NOTES

1. David H. Petraeus, "Learning Counterinsurgency: Observations from Soldiering in Iraq," *Military Review* 86, no. 1 (January–February 2006): 8.

2. In fact, the pig story might be more closely associated with the American style of PSYOP than even I might have imagined at the time of my newsstand encounter. After relaying this story at a conference at Texas Tech University in October 2006, distinguished former OSS agent Rene Defourneau excitedly approached the podium from where I spoke. He asked me if I knew that "General Pershing did use pigs in the Philippines . . . and it worked!"

3. Defense Science Board, Report of the Defense Science Board Task Force on Strategic Communication, September 2004, 12 (emphasis in original).

4. Joint Chiefs of Staff, *Information Operations*, Joint Publication 3–13, February 13, 2006, II-1.

5. Kenneth Payne, "The Media as an Instrument of War," *Parameters* 35, no. 1 (spring 2005): 81.

6. Organization and Functions Psychological Operations, July 10, 1968, Folder 17, Box 13, Douglas Pike Collection: Unit 03—Insurgency Warfare, The Vietnam Archive, Texas Tech University, Lubbock, Texas (VA-TTU).

7. Paper delivered at the National Strategy Information Center Conference by Barry Zorthian, "The Use of Psychological Operations in Combatting 'Wars of National Liberation,'" March 11–14, 1971, "Material on JUSPAO," Papers of Barry Zorthian, 4A39, 40, Lyndon Baines Johnson Library, Austin, Texas (LBJL). JUSPAO

came under the supervision of United States Military Assistance Command, Vietnam (MACV) with the creation of Civilian Operations and Revolutionary Development Support (CORDS) in May 1967.

8. Remarks, Dinner Meeting of the Texas Electric Cooperatives, Inc., May 4, 1965, John Woolley and Gerhard Peters, The American Presidency Project [online]. Santa Barbara, CA: University of California (hosted), Gerhard Peters (database) <http://www.presidency.ucsb.edu/ws/?pid=26942> (October 10, 2006).

9. Russell F. Weigley, *The American Way of War: A History of United States Military Strategy and Policy* (New York: Macmillan, 1973).

10. Russell F. Weigley, "How Americans Wage War: The Evolution of National Strategy," in John Whiteclay Chambers II and G. Kurt Piehler, eds., *Major Problems in American Military History* (Boston and New York: Houghton Mifflin Company, 1999), 2–6.

11. U.S. Army Major General Peter W. Chiarelli and Major Patrick R. Michaelis, "Winning the Peace: The Requirement for Full Spectrum Operations," *Military Review* 85, no. 4 (July–August 2005): 4.

12. Curtis D. Boyd, "Army IO is PSYOP: Influencing More With Less," *Military Review* 87, no. 3 (May–June 2007): 72.

13. U.S. Congress. The Honorable Edward P. Djerejian, Chairman, Advisory Group on Public Diplomacy for the Arab and Muslim World. Testimony Before the Subcommittee on the Departments of Commerce, Justice and State, the Judiciary and Related Agencies, February 4, 2004.

14. "Rumsfeld: U.S. Failing in Battle of Ideas," *The Philadelphia Inquirer*, March 24, 2006, 1.

15. Jerrold M. Post, "Psychological Operations and Counterterrorism," *Joint Forces Quarterly* no. 37 (April 2005): 105.

16. Report, U.S. Army Major Joseph L. Cox, *Information Operations in Operations Enduring Freedom and Iraqi Freedom: What Went Wrong?* May 25, 2006, School of Advanced Military Studies, United States Army Command and General Staff College, Fort Leavenworth, Kansas.

17. Boyd, "Army IO is PSYOP," 72.

18. Cox, *Information Operations*.

19. U.S. Army Captains Stoney Trent and James L. Doty III, "Marketing: An Overlooked Aspect of Information Operations," *Military Review* 85, no. 3 (July–August 2005): 70–74.

20. Kathleen T. Rhem, "Burning Afghan Bodies Resulted from Poor Judgment, Not War Crimes," *American Forces Information Service*, November 29, 2005 <http://www.globalsecurity.org/military/library/news/2005/11/mil-051129-afps04.htm> (May 19, 2007).

21. J. D. Scanlon, "Truth: In Defense of Military Public Affairs Doctrine," *Military Review* 87, no. 3 (May–June 2007): 92–96.

Chapter One

Blurring the Line

OVERVIEW

This chapter is divided into three sections. The first, "A Matter of Hearts and Minds," demonstrates that before the onset of Cold War between the United States and its communist adversaries, American officials managed to maintain some distinction between the efforts they made to communicate informatively and those they initiated to speak persuasively. Afterwards, beginning in Vietnam, policy makers came to identify all efforts that the United States made to communicate, whether persuasive or informative, under the phrase of psychological operations.

The second section, "We Don't Expect Miracles," underscores the tendency of Americans to scapegoat the Vietnamese as the reason for the failure of their PSYOP program in Vietnam. During the war and since, both participants and scholars have cited an unbridgeable cultural gap that faced Americans in Vietnam, but they have ignored the unwillingness of American officials to cross it.

"By the Numbers," this chapter's third part, suggests that despite their repeated insistence otherwise, American officials exhibited little real interest in utilizing their PSYOP program to fulfill its stated purpose: to establish a genuine dialogue with Vietnamese. Instead, United States leaders focused on their program's ability to secure improved public relations in Vietnam, at home and around the world.

A MATTER OF HEARTS AND MINDS

This study examines an area of America's involvement in Vietnam that scholars have failed to scrutinize closely, the psychological operations program

(PSYOP) carried out by the United States Information Agency (USIA) through the Joint United States Public Affairs Office (JUSPAO) to win the "hearts and minds," of both Vietnamese and others, for the support of American policy in Southeast Asia. The historiography of America's involvement in Vietnam, where it is at all concerned with the effort made by United States officials to gain support for their actions, manages to provide this crucial endeavor with only a superficial treatment, moving little beyond the testimony of American officials about the program's perceived effectiveness.[1]

This work connects America's use of PSYOP with the nation's history of using information programs and psychological warfare (PSYWAR) to communicate its policies throughout the twentieth century. Moreover, it examines the effort within the context of America's overall attempt to buttress the Republic of Vietnam (RVN), the government in Saigon that the United States endorsed as a necessary Vietnamese bulwark against communist expansion in Southeast Asia. This work illustrates that in Vietnam, American officials lost sight of the line that their predecessors had attempted to maintain between United States information and PSYWAR. Beginning with their experience during the First World War, American leaders considered PSYOP to contain distinctly different elements, information, propaganda than PSYWAR. While commentators utilized these terms almost interchangeably throughout the twentieth century, government officials consistently strove to create more precise definitions to guide their initiatives.[2]

Americans construed their information programs as ways to "facilitate the development of an informed and intelligent understanding, at home and abroad, of the status and progress of the activities and aims of the government." Most important, they deemed that these efforts required officials to "rule out the technique of falsehood and much, if not all, of the strategy of terror." Officials developed this strategy out of faith that "if enough people abroad, both our friends and our enemies, are shown the truth about this country, it will shorten the war."[3] They also acted out of necessity. During both world wars and in the Cold War that followed, the United States had accused its enemies of disseminating propaganda to aid in the extension of totalitarian regimes, whether imperialist, fascist, or communist. Consequently, Americans remained "suspicious even of the term public relations" regarding the activities of their nation in general and of their chief executive in particular.[4]

The expression "public information" resulted from the American effort to demonstrate the difference between their nation and its enemies, both for themselves and the citizens of other countries they sought to reach with their message. Americans staffed their information programs with journalists, educators, and librarians, all members of professions that value "impartiality" as a means to achieve credibility with others. They turned to the members of these fields to provide their agencies with direction as well, often selecting well-known journalists such as Elmer Davis and Edward R. Murrow to provide these efforts with a credible public face. American officials cited their willingness to "let the facts speak for themselves" through their information agencies, confident that foreign groups would "learn about and understand the United States and our motives and actions."[5]

Americans defined their information programs according to the beliefs they held about themselves and about the rest of the world. They considered the United States as uniquely capable of assuming an objective and selfless perspective on world affairs. According to George Venable Allen, USIA director (1957–1960), the notion became "prevalent among people in all walks of life that all we need to do is explain ourselves, our policies, and our way of life to foreign peoples and they will love us." Allen called it an "understandable point of view" that sprung from "an admirable characteristic of Americans"; their knowledge that they are not "war-mongering imperialists," but "good, kind-hearted people who have no designs on anyone else's territory and that if foreigners could only know us as we are, they would like us."[6]

By the time the presidential administration of Dwight David Eisenhower created the USIA in 1953 to combat the "hostile intentions" of America's Cold War adversary the Soviet Union, information officials considered their activities on behalf of United States policies to mimic the efforts of an independent news organization.[7] USIA director Theodore Streibert (1953–1956) assured President Eisenhower soon after taking the reins of the new agency that it would "concentrate on objective, factual news reporting."[8] In outlining his operation's strategic principles, he shunned for America the "purely manipulative approach of totalitarian propaganda." The United States, Streibert insisted, "must be content with the facts," and that "the facts will in the long run be the most effective arguments for our national interests."[9]

Conversely, American officials came to conceptualize PSYWAR, and often propaganda, in terms justifying their nation's need to "steal the ball and kill the umpire" in order to engage totalitarian regimes on equal footing.[10] United States leaders confined their nation's use of this subversive activity to the battlefield, and they used it primarily in hopes of immediately increasing the number of enemy defections. While information programs relied on the news, conveyed through print, radio, and film that officials openly attributed to the United States, the chosen instruments of psychological warriors tended toward leaflets, loudspeakers, and various forms of trickery that often concealed any American origin. Information officials appealed to reason, but psychological warriors strove to strike an emotional chord in weary soldiers during the heat of battle. American information programs reached out for the civilian's mind as PSYWAR grasped for the enemy soldier's heart.

This study notes that through the two world wars, PSYWAR and information remained different activities from one another in the ways American officials conceptualized and carried them out. In the Cold War world where United States leaders faced an increasingly difficult task in trying to discern the front lines from the rear areas, however, earlier distinctions that officials made between information and PSYWAR no longer appeared valid. The information program in the 1950s, first accused by critics of sympathizing with communists and then subjected to charges of losing ground in the Third World, proved increasingly willing to blur the differences that Americans had historically observed between PSYWAR and information. The JUSPAO formally completed the process in Vietnam by integrating the two activities, serving as the first agency in American history to carry out simultaneously both PSYWAR and information campaigns under the umbrella of PSYOP.

At the time, officials hailed the new union that the JUSPAO constituted as representing a "revolutionary development" in the realm of United States psychological action. They cited its inventiveness as born from necessity because they believed that "The war in Vietnam is many wars, with no frontiers."[11] Nevertheless, by combining the functions of PSYWAR and information into one agency, they created insurmountable obstacles for themselves that they failed to anticipate and proved unable to overcome in Vietnam. The situation only worsened as officials insisted on continuing to justify their new agency's activities according to the traditional refrain of objectivity that had accompanied earlier information programs. While the JUSPAO provided material endorsed by the United States Mission as official war information to domestic reporters as well as foreign scribes, the agency immediately faced credibility issues that arose from its dual role encompassing PSYWAR and disseminating information.

The daily conferences that agency representatives held in Saigon came to be dubbed by the skeptical reporters who attended them as the "Five O'Clock Follies." The problem only became exacerbated as the war went on and the gap widened between the situation detected by journalists outside and the material officials fed them inside of the pressroom.[12] *Time* magazine reported that "Frequently, information provided has been found to be shaded. Often it is incomplete. Sometimes, although not as often as critics allege, it is deliberately misleading." The newsweekly labeled the whole ordeal reporters experienced in Vietnam while attempting to do their jobs as "exasperating."[13] Such allegations, even with the care this one demonstrated not to overstate the case, did little but add to the atmosphere of mistrust that enshrouded Americans who covered the war and those who carried it out.

Rather than reassessing the situation, JUSPAO officials became more defensive about the arrangement to carry out the new style of PSYOP that the

United States had established. Agency director Barry Zorthian (1965–1968) willingly acknowledged "that much of what we are doing here is admittedly not USIA's basic mission," but he justified the integration of PSYWAR and information as a "built-in characteristic of the JUSPAO concept." He questioned whether there existed in Vietnam any "difference between information and psychological warfare" and wondered if even "such a distinction is valid outside of the classroom." Zorthian contended, "On the scene, an artificial line between psychological warfare and information does not stand up to examination."[14]

In assigning the JUSPAO its dual role in Vietnam, American officials failed to note the lessons learned by their predecessors who took great care during both the First and the Second World Wars to keep their information programs distinct from their PSYWAR campaigns. For example, William J. Donovan, while serving as the director of the Office of the Coordinator of Information (1941–1942), noted America's need as a democratic nation with a history of a free press to avoid "creating a tie-up between domestic and foreign propaganda agencies." Otherwise, he believed a situation would manifest that "would compromise or destroy the nation's security, and impair the effectiveness of psychological warfare as a weapon."

Donovan understood that as a democratic country with a history of an independently operated press that acted as a government "watchdog," the United States needed "to fix an arbitrary line between domestic propaganda and foreign psychological warfare."[15] It needed to make separate, if even in appearance, the initiatives it authorized to maintain support and morale at home from the efforts it took overseas to break the will of America's enemies. Failing to do so meant risking the trust of an American public sensitive to being subjected to propaganda campaigns carried out by its own government, one ostensibly committed to the notion of a free press. This insight fully eluded American officials regarding their nation's PSYOP policies in Vietnam. They repeatedly deflected criticism concerning the arrangement by reciting the USIA creed of objectivity. "The fact is," officials frequently asserted, "that USIA doesn't have one version of the truth to give to foreign newsmen and another to give to Americans."[16]

In his work that chronicles the development of the USIA from 1953 through the 1960s when he served as the agency's Deputy Director, Thomas C. Sorensen, asserts that "there is not necessarily a conflict between veracity and advocacy, that it is possible to be both truthful and persuasive." The United States "has not always been persuasive," Sorensen writes, "but it has always tried to be truthful." The former USIA official contends that the problems faced by the USIA in the 1960s, both at home and abroad, resulted from "the double standard applied to the conduct of nations." He explains: "More is expected and demanded of the United States than of other nations." Consequently, he argues, "the pickets and paint-splashers" harassed American politicians, but they made "no complaint" when "Hanoi and the Vietcong launched their campaign of terror and subversion in South Vietnam."[17]

This point is a valid one, but it neglects to consider the perspective held by those on the receiving end of the message that USIA delivered. William J. Donovan grasped during World War II that it did not matter to what extent truthfulness and persuasion blurred in the eyes of officials, the American public needed reassurance that its government officials could be trusted. Leaders needed to draw a line, visible to American citizens at all times, that the agency charged with disseminating information about the war at home differed from the efforts made by the United States to break the will of its enemies overseas. In Vietnam, Americans officials failed to do so. Leaders gave the JUSPAO an unprecedented authority to communicate with all of the audiences Americans sought to reach in Vietnam, in the United States, and around the world. Simultaneously, the JUSPAO acted as propagandist, government news agent, and psychological warrior. As a result, discrepancies between the information endorsed by the JUSPAO and the story conveyed by other sources, no matter how slight, immediately raised damaging questions regarding the intentions of American leaders.

WE DON'T EXPECT MIRACLES

This work surveys the most thorough of the few existing accounts of America's PSYOP in Vietnam.[18] While acknowledging the important contributions made by these studies, particularly their descriptions of its operations, it calls attention to the superficial analyses they offer readers. Written by those individuals who participated in the USIA program, these narratives fail to stray far from the agency's official explanation of events in Vietnam, one that remained rooted in the fixed perceptions about Vietnamese, and about the USIA mission, that Americans held during the war. They identify scapegoats, but they sacrifice understanding in the process. These works do little to examine the fundamental assumptions Americans took with them to Vietnam, but they do reveal that the United States, contrary to what critics of America's war effort contend, placed heavy emphasis on cultural considerations.

In fact, as opposed to ignoring culture, this study contends that Americans put Vietnamese into a well defined and tightly sealed cultural box from where any escape proved impossible. This time, officials took their cue from plans Americans had devised for use during the Second World War in the event of a United States invasion of Japan. They demonstrated a proclivity to objectify Vietnamese people by stereotyping them as products of a static and primitive culture, removed from America by not only space but also by time.[19] The USIA program did little to alter this view, despite its allegedly assigned role of opening up channels of communication to develop rapport among Americans, RVN officials, and ordinary Vietnamese.

The material that the USIA compiled to inform American officials about Vietnamese society amounted to gross generalizations that fed the mentality already existent among Americans generally that people in Vietnam, consistent with Asians in general, constituted an unknowable, listless, and considerably less developed "other."[20] The JUSPAO, for example, indicated that "the underlying measure" of an American's success in Vietnam consisted of the "degree to which he works himself out of a job" by bringing along his counterpart "to the point where the Vietnamese can carry on without assistance." No small task, the American agency replied, citing that "progress" in Vietnam transpired "not in a continuous curve, but in pulses."[21]

The most unfortunate part of the story is that on occasion American PSYOP personnel revealed the promise of their program to enable a more subtle understanding of modern Vietnam than what United States leaders ultimately privileged in their analysis of the war. As historian Gabriel Koko observes, "The American government had capable people working on Vietnam and they knew a great deal about that nation."[22] Yet, insightful advice often went unheeded. Operating in the field, for example, some JUSPAO representatives discerned the difference between the Vietnamese they encountered and the stereotypes that Americans had already constructed. They began to separate Western idealizations of "oriental ways" from the historical and modern realities of Vietnamese life. They questioned the compatibility of American initiatives with Vietnamese interests and suggested that officials gear their programs to the popular concerns they detected. The JUSPAO field representative for Phuoc Tuy and Vung Tau Provinces reported during the summer of 1966 that "our conversation with people, generally, would seem to indicate little or no change in attitudes toward the GVN [Saigon government] despite pacification process." He indicated that, while "the vast majority of people in Phuoc Tuy are pro-GVN in the sense that they prefer GVN control to VC [insurgents] control," it did not mean "they believe the GVN will bring them the kind of life they desire."[23]

The report makes clear the pervasiveness of the view held by the people living in these two provinces, not far from Saigon and ostensibly well under RVN control, that the officials endorsed by the United States possessed little interest in their welfare. The report quotes one village elder in the Vung Tau hamlet who it suggests sums up the popular view. "People have lost faith in the government because it has failed to control the black-marketeers that are the chief cause of high prices." It also cites that people held minimal hope that the democratic process that American officials promised for Vietnam's future would provoke much change for the better. The report identifies that "many" suspected "the elections will not be fair and that candidates hand-picked by the Government will win."[24]

BY THE NUMBERS

This study identifies the political subculture that spawned America's PSYOP policies in Vietnam. Ideas held by American officials regarding their nation's perspective in the Cold War, and its privileged place in the world order, enabled them to create a program that too often elevated form over substance for the sake of appearance and in the interest of national prestige. American officials often proved too willing to abide by a "soft lights and trick focus" mentality that grew ever more pervasive as they became more defensive about the increasing level of criticism regarding their nation's position.

Pressured for results that could demonstrate tangible progress for the leaders in Washington who demanded them, JUSPAO carried out an array of programs aimed more at garnering good press and generating photo opportunities than at genuinely winning the hearts and minds Americans professed to covet for the RVN. The psychological operation's effort, like the military's, became reduced to a numbers game in which quantities of leaflets dropped and hours of broadcasts programmed superseded any efforts made to listen and respond to popular concerns. One JUSPAO representative bragged to *Time* magazine in 1965 about his agency's 3.5 million leaflet per week effort. "We're the world's worst litterbugs."[25]

In fact, American officials related their success in "winning hearts and minds" almost exclusively in terms of the amount of product that the JUSPAO delivered, with little thought given to its qualitative value for the people it presumably served to uplift in Vietnam. USIA director Leonard H. Marks (1965–1968) identified "one of the most immediate and tangible yardsticks of JUSPAO contribution to pacification" as the number of agency personnel "deployed for provincial operation in South Viet-Nam." He noted with great satisfaction that by January 1966, 31 JUSPAO officers "were stationed and functioning in provincial locations throughout the country." For Marks and other American officials, "increased JUSPAO capabilities" equated to success in Vietnam. The USIA chief proposed the jump "in offset printing in Saigon since July 1965," which rose by almost 10 million impressions per month by March 1966, "may provide a graphic picture of progress made by JUSPAO."[26]

The PSYOP program that Americans created in Vietnam operated in a sphere separate from the war they pursued. Psychological operators relied on communicating with Vietnamese by drawing their appeals from the anachronistic version of Vietnam's culture anthropologists had created, rather than making a concerted effort to understand contemporary Vietnamese concerns. Americans generated products devised according to the stereotypes they harbored about Vietnamese as a primitive people motivated primarily by their "irrational" belief systems. They bombarded their enemies in the field with "weird, electronic cacophonies" that aimed to "raise terrifying images of forest demons" among combatants they perceived as "superstitious terrorists." Meanwhile, to rally Vietnamese to support the RVN, they dispatched "rural spirit" drama troupes to villages that set about "maligning Red China and Ho Chi Minh" by interspersing "propaganda skits" with "classical Vietnamese ballads" and "boogie-woogie."[27]

American officials showcased the JUSPAO effort for observers while remaining oblivious to how contrary to their nation's actions elsewhere in the war these activities must have appeared to individuals paying at all attention. The situation became something of an absurdist "bread and circus" routine as the information program offered up radio shows, television programs, and drama teams, while American military might depopulated villages and destroyed arable land. As the "boogie-woogie" troupes made their rounds in the fall of 1965, for example, the American military effort under United States Army General William Westmoreland began to implement the strategy of "territorial denial," a concerted effort to render as "uninhabitable" rural areas that Americans deemed as "uncontrollable." Increasingly larger swaths of territory designated as "free fire zones" became subject to "every weapon" in the United States "arsenal to separate the remaining Vietnamese rural population from the insurgent NLF [National Liberation Front]."[28]

The PSYOP program, operating in the midst of such indiscriminate violence and upheaval, came to represent an opportunity for officials to demonstrate tactical success stories for the rest of the world to witness how the "other war," the one Americans waged for hearts and minds, progressed. It did little, however, to bring about more lasting reforms that might have furthered a strategy to produce a more widespread popular base of support for the RVN than what it possessed.

Officials failed to capitalize on the program's promise to empower Vietnamese by giving them a means whereby their voices might be heard and where their concerns about a modern Vietnam might be met. Instead, America's leaders used PSYOP with the expectation they might mollify a Vietnamese people they considered apathetic with promises of delivering an improved standard of living and a tutorial in American democratic values. The rhetoric Americans instilled in the material the USIA churned out could not have more starkly contrasted to the pacification initiatives that their nation simultaneously conducted.

America's PSYOP program actively enabled the way Americans waged war and conducted nation building in Vietnam by generating and maintaining a simplistic portrait of Vietnam despite its unique position and assigned task to do otherwise. It fostered the idea among Americans that Vietnamese did not yet stand capable of responsibly governing themselves, and it blamed setbacks on what officials simply dismissed as the Vietnamese way of doing things. Under immense pressure to bring the stability to Vietnam that Americans wholeheartedly believed Vietnamese could not bring about themselves, Vietnam's people became little more than props in the eyes of officials who expected them to animate American ideas.

Vietnamese perspectives on matters went unsolicited and generally undetected as Americans exhibited tendencies toward paternalism at best and colonialism at worst. Through all of the rhetoric they offered about democracy, United States officials never managed convincingly to demonstrate any faith in the ability of Vietnamese to determine their own path. The actions they took, and failed to take, communicated more clearly than all the leaflets and broadcasts combined that their information program generated.

The most significant contribution made by the PSYOP program to the American effort, this study proposes, involved aiding and abetting the refusal of American policy makers to reconsider their fundamental assumptions regarding Vietnamese.

The USIA, either directly or through the JUSPAO, operated in Vietnam for two decades, yet it never developed a more sophisticated understanding of the nation or its people than what it initially offered. Instead, it reflected and made more rigid existing American ideas that provided a comfortable and convenient excuse for officials who cited culture and not their own contributions to policy as the biggest obstruction the United States faced in Vietnam. Owing largely to their own intransigence, as Central Intelligence (CIA) historian Thomas L. Ahern observes, "US policymakers and program managers did in fact fail to comprehend what they were up against, both when they tried to help Saigon win the allegiance of the peasantry and when they tried to destroy the Viet Cong."[29]

NOTES

1. The only works wholly concerned with the American psychological program in Vietnam are both memoirs that lack archival material and describe the activities of the

Joint United States Public Affairs Office (1965–1972) while providing minimal analysis. Robert W. Chandler, *War of Ideas: The U.S. Propaganda Campaign in Vietnam* (Boulder, Colorado: Westview Press, 1981); and Harry D. Latimer, *US Psychological Operations in Vietnam* (Providence, Rhode Island: Brown University Press 1973).

2. The United States Department of the Army officially adopted the phrase of "psychological operations" as the all-inclusive label for America's psychological activities in 1971. This study utilizes the terms "information," "propaganda" and "psychological warfare," as American officials did through the course of United States history. That is, this work uses "information" to describe efforts made by the United States to gain support for long-term national objectives. It employs "propaganda" to note American attempts at influencing thinking on immediate issues of the day, and it identifies the nation's initiatives on the battlefield to undermine enemy morale as "psychological warfare." Overall, this work argues that through the Second World War American officials considered these activities as distinct from one another and, at times, even as unrelated endeavors. During the Cold War, though, policy makers came to view propaganda, psychological warfare, information, and a program of cultural exchange as overlapping subsets of an integrated national program of psychological operations. For evolution of terminology, see William E. Daugherty, "Origin of PSYOP Terminology," in *The Art and Science of Psychological Operations: Case Studies of Military Application*, proj. dir. Daniel C. Pollock, ed. Ronald De McLaurin, Carl F. Rosenthal, Sarah A. Skillings, and others, 2 volumes (Washington, D.C.: American Institutes for Research, 1976), 18–19; Theodore C. Sorensen, *The Word War: The Story of American Propaganda* (New York, Evanston, and London: Harper & Row, 1968), 2–5.

3. Joseph Barnes, "Office of War Information," in *America's Weapons of Psychological Warfare*, Robert E. Summers (New York: Arno Press, A New York Times Company, 1972), 21.

4. Francis M. Watson Jr., "PSYOP and Related Activities," in *The Art and Science of Psychological Operations*, I: 57.

5. Raymond J. Barrett, "PSYOP: What Is It? What Should We Do About It?" *The PSYOP Digest* 1, no. 1 (Jan–Mar 1977): 29.

6. George V. Allen, "What the U.S. Information Program Cannot Do," in *Propaganda and the Cold War: A Princeton University Symposium*, ed. John Boardman Whitton (Washington, D.C.: Public Affairs Press, 1963).

7. "U.S. Information Agency: Strategic Principles," undated, *Foreign Relations of the United States* (*FRUS*), *1952–1954*, II: 1763.

8. "The Director of the United States Information Agency (Streibert) to the President (Eisenhower)," October 27, 1953, *FRUS, 1952–1954*, II: 1755.

9. "U.S. Information Agency: Strategic Principles," undated, *FRUS, 1952–1954*, II: 1763.

10. Wallace Deuel, *History of OSS* (5 volumes, undated, unpublished), I, 105, box 122A, William J. Donovan Papers, United States Army Military History Institute, Carlisle, Pennsylvania (MHI).

11. Booklet, *JUSPAO Vietnam: General Briefing Book*, January 1966, "Vietnam Kit," box 426, Office Files of Fred Panzer, LBJL.

12. See Daniel Hallin, *The "Uncensored War": The Media and Vietnam* (New York: Oxford University Press, 1986); William M. Hammond, *Public Affairs: The Military and the Media, 1962–1968,* United States Army in Vietnam, David F. Trask, general editor (Washington, D.C.: Center of Military History, United States Army, 1988); Phillip Knightley, *The First Casualty: From the Crimea to Vietnam: The War Correspondent as Hero, Propagandist, and Myth Maker* (New York and London: Harcourt Brace Jovanovich, 1975); John Mecklin, *Mission in Torment: An Intimate Account of the U.S. Role in Vietnam* (New York: Doubleday, 1965); Caroline Page, *U.S. Official Propaganda During the Vietnam War, 1965–1973: The Limits of Persuasion* (London and New York: Leicester University Press, 1996).

13. "Facts of Life in Vietnam," *Time* magazine, May 7, 1965, 13.

14. Letter, to Leonard Marks (USIA Director) from Barry Zorthian (JUSPAO Director), December 21, 1966, "Vietnam Correspondence," National Security Files (NSF), Files of Robert W. Komer (1966–1967), LBJL.

15. Letter, William J. Donovan to Brigadier General William B. Smith, March 23, 1942, *Records of the JCS, Part 1: 1942–1945, Strategic Issues* (Frederick, Maryland: University Publications of America, 1983), XI, F246, microfilm.

16. Memo, to President Johnson from USIA director Carl T. Rowan, May 12, 1965, "USIA Vol. 4, 4/14/65, 2 of 2," NSF-USIA File, box 74, LBJL.

17. Sorensen, *The Word War*, 4.

18. Chandler, *War of Ideas*; Latimer, *US Psychological Operations in Vietnam*; Carl T. Rowan, *Breaking Barriers: A Memoir* (Boston, Toronto, London: Little, Brown and Company, 1991); Sorensen, *The Word War*; Barry Zorthian, "The Use of Psychological Operations," in *Lessons of Vietnam: Three Interpretive Essays, ed.* Dennis J. Duncanson, Richard A. Yudkin, and Barry Zorthian (New York: American-Asian Educational Exchange, 1971).

19. Johannes Fabian, *Time and the Other: How Anthropology Makes Its Object* (New York: Columbia University Press, 1983); David Spurr, *The Rhetoric of Empire: Colonial Discourse in Journalism, Travel Writing, and Imperial Administration* (Durham, North Carolina: Duke University Press, 1993); Edward Said, *Culture and Imperialism* (New York: Vintage Books, 1993); Nicholas Thomas, *Out of Time: History and Evolution in Anthropological Discourse* (Cambridge: Cambridge University Press, 1989).

20. Mark Philip Bradley, *Imagining Vietnam and America: The Making of Postcolonial Vietnam, 1919–1950* (Chapel Hill and London: University of North Carolina Press, 2000); Matthew Frye Jacobson, *Barbarian Virtues: The United States Encounters Foreign Peoples at Home and Abroad, 1876–1917* (New York: Hill and Wang, 2000).

21. Booklet, *JUSPAO Vietnam*, 8.

22. Gabriel Kolko, "Familiar Foreign Policy and Familiar Wars: Vietnam, Iraq . . . Before and After," in *Iraq and the Lessons of Vietnam: Or, How Not to Learn From the Past*, ed. Lloyd C. Gardner and Marilyn B. Young (New York and London: The New Press, 2007), 163.

23. Report, PHUOC TUY/VUNG TAU PROVINCES, undated, "Reports of JUSPAO Field Representatives for the month ending August 31, 1966," September 20, 1966, Counterinsurgency File (CI), Richard M. Lee Papers, MHI.

24. Report, PHUOC TUY/VUNG TAU PROVINCES, undated.
25. "Psywar," *Time* magazine, October 29, 1965, 36.
26. Progress Report on JUSPAO Support to Pacification, June 3, 1966, "Psy War," box 21, NSF—Komer—Leonhart File, LBJL.
27. Progress Report on JUSPAO Support to Pacification, June 3, 1966.
28. George McT. Kahin, *Intervention: How America Became Involved in Vietnam* (New York: Alfred A. Knopf, 1986), 406.
29. Thomas L. Ahern Jr., *CIA and Rural Pacification in South Vietnam*, Central Intelligence Agency (CIA) History Staff, Center for the Study of Intelligence, undated, 393.

Chapter Two

Truth or Consequences

OVERVIEW

This chapter is divided into three parts. The first, entitled "The Whole Process of Communication," provides an overview of the program that the United States carried out in Vietnam through the USIA and the JUSPAO. It illustrates the extensiveness of the operation and cites the rhetorical support offered PSYOP by American officials in Saigon and Washington. This chapter's second section, "Neglecting Hearts and Minds," makes apparent the historiographical lacuna that exists concerning PSYOP in Vietnam. It suggests that while a number of reasons explain scholars' neglect of the USIA in Vietnam, including the difficulty involved in demonstrating the agency's effectiveness, the oversight mainly stems from the location of PSYOP outside the parameters of traditional military and diplomatic histories of the war.

"Making Bad Policy Palatable," this chapter's third part, proposes that whatever tactical successes the USIA accomplished through the JUSPAO operation, the agency failed to deliver on its conceptual promise. Officials carried out American PSYOP in Vietnam with little interest in opening a multidirectional dialogue. To the extent the agency informed American policy, it did so by reinforcing the "oriental" stereotype Americans possessed that cast Vietnamese as primitive, passive, and superstitious. Otherwise, the JUSPAO and the USIA contributed to the American effort by serving officials as the very things that Americans historically have professed to abhor, propaganda departments.

Any advantages that officials hoped to gain in the "war for hearts and minds" by using these agencies clearly became eclipsed by the costs they incurred by providing critics of the American effort with legitimate questions concerning United States credibility. This section also makes clear the need to reconsider the American PSYOP program in Vietnam because of parallels apparent between the USIA program and the one currently employed by the United States in support of its "nation building" efforts in places such as Afghanistan and Iraq.

THE WHOLE PROCESS OF COMMUNICATION

In Vietnam the USIA worked to support the American-backed Republic of Vietnam (RVN) from its creation in the 1950s to its demise in the 1970s. The agency carried out a variety of activities to buttress the legitimacy of RVN rule over Vietnam's south and to win support for America's involvement in Vietnamese affairs. It advised its RVN counterpart, the Vietnamese Information Service (VIS), managed relations between the United States Embassy and the press, and opened libraries to enable Vietnamese to become familiar with American history, government, and culture. The agency also strove to erode the credibility of Vietnamese communists as it devised campaigns to disparage both the Democratic Republic of Vietnam (DRV) in the north and the insurgent National Liberation Front (NLF) in the south.

As the United States steadily increased its support of the RVN in the 1960s, the role played by the USIA expanded as well. The agency received authority from Washington to supervise all PSYOP in Vietnam through the JUSPAO, an office created by the Lyndon Baines Johnson presidential administration in 1965. The JUSPAO, directed by the former USIA Public Affairs Officer (PAO) in Saigon, drew its personnel from the USIA, the United States Agency for International Development (USAID), and the United States Military Assistance Command Vietnam (USMACV). It employed six hundred and ninety-five people and operated on an annual budget of ten million dollars. Its staff of two hundred and forty-five Americans, including one hundred and sixteen from the military.

Except for a few personnel drawn from USAID who provided JUSPAO with technical support, the USIA met the rest of the agency's personnel needs by siphoning off ten percent of its human resources deployed elsewhere. The new agency represented an unprecedented combination of American civilians and soldiers under one roof that received support from Vietnamese employed by the JUSPAO as translators, survey takers, and field support.[1]

In the words of its director from 1965–1968, Barry Zorthian, a former marine and reporter for *Time* magazine, the responsibilities of JUSPAO encompassed "nothing more—and nothing less—than the whole process of communications in a wartime situation."[2] The JUSPAO served as the medium for American officials to make their case about Vietnam as persuasively as possible both to the rest of the world and to their own nation. It conveyed the official view of the United States government to "the domestic audience, the friendly alien audience, the hostile alien audience, the assisting nation audience, the military, the civilian, the bureaucracy, the peasant, the urban dweller—all as required."[3]

Zorthian hardly overstated the jurisdiction of his agency, as leaders in Washington had pledged to make psychological considerations "part and parcel of every step [the United States] took in Vietnam."[4] So too did American officials in Saigon who witnessed events that routinely underscored their nation's need to grasp better the climate of opinion in Vietnam and act on it. In 1966, the United States Ambassador to Vietnam, Henry Cabot Lodge, acted as the awards presenter at a ceremony arranged by the USIA. He personally appeared to honor its outstanding officers and to proclaim its operations as instrumental to America's effort. The event did not fully proceed as planned, though. According to the *USIA Correspondent*, an internal publication that the USIA published monthly to inform its personnel of agency activities around the world, the affair came to be "marked by both tragedy and accomplishment."[5]

The day had originally been intended by American officials to call attention to the performance of the JUSPAO by citing the work of individuals such as Frank Scotton, a USIA field representative in Vietnam since 1962. Scotton stood to be recognized for his marked contribution to what the agency deemed as "the development of an imaginative counter-insurgency information program."[6] The celebration orchestrated for Scotten and the others that the USIA had chosen to fête for their outstanding service had become bittersweet as a result of what transpired only the day before the scheduled activities took place. JUSPAO field officer Nguyen Van Thuy had taken a jeep-load of Vietnamese Information Service (VIS) personnel on an inspection trip of communication facilities in Hau Nghia province. About 35 miles northwest of Saigon, the group found itself under attack by insurgents who opposed RVN rule. The ambush left Nguyen and five of the six others who traveled with him dead, and the incident cast a dark cloud over the occasion staged the next day in Saigon.

As happened often in Vietnam, motivating the living while honoring the dead now went hand in hand. Thanking his staff for their continuing good work, JUSPAO director Barry Zorthian remarked, "like Mr. Thuy, many of our people here daily perform outstanding work under adverse conditions and great personal danger." When it came time to fulfill the day's original agenda and present the awards earned by Scotten and his colleagues, Ambassador Lodge addressed the need to draw lessons from the previous day's unfortunate incident. He stressed, "One of the basic needs here is to know what the people think, why they think it, and how to communicate with them. Without this knowledge, which is the basis of JUSPAO, our work here would be lost."[7]

The contributions made by PSYOP personnel also did not escape the attention of observers on Capitol Hill. In the spring of 1966, Utah's democratic congressperson David S. King touted the information program's merits before his colleagues. On the floor of the United States House of Representatives, King pointed out:

> We hear so much about the conventional battles that are raging in Vietnam, and about the statistics on casualties suffered, lives lost, and planes and materiel destroyed. The real war however, is being fought, not for the bodies but for the minds of the Vietcong. The work of JUSPAO [Joint United States Public Affairs Office] is no less significant because it receives so little recognition. In my opinion it is our ultimate weapon.[8]

King wanted to make part of the public record what he deemed as the "splendid work of our JUSPAO organization and the success it has had in the field of psychological warfare." He echoed the Johnson administration's belief that the media had become preoccupied with covering initiatives that went poorly for America in Vietnam, and he wished to identify something about "what we are doing right." He asserted that PSYOP had not received sufficient acclaim from Americans because its work did not lend itself to splashy headlines in the newspapers. Whatever mention the JUSPAO program garnered too often only consisted of reports occupying the "backside of the news," and King lamented that such treatment grossly underestimated the important role it played in the outcome of America's Vietnam policies. "Just as crucial as, and in the long run more important than, the current military effort in Vietnam," he insisted, "is the struggle for the hearts and minds of the Vietnamese."[9]

Congressperson King, largely in the interest of building support for the policies of his political party's occupant of the White House, provided his listeners with a decidedly partisan account of JUSPAO success. His political motivations aside, King's plea to recognize the importance of JUSPAO programs to American ambitions in Vietnam echoed observations made by others. Observers of the war, both within and outside of America's government, identified the conflict's affective dimension as containing an essential key for American success, and they believed the JUSPAO program could secure it. They had apparently provided PSYOP personnel with a prominent role to play in carrying out America's Vietnam policy.

NEGLECTING HEARTS AND MINDS

Despite its apparent prominence in the minds of national leaders during such a noteworthy time in American history, the JUSPAO received little attention from contemporary observers, and it has garnered even less scrutiny by scholars since. It is nearly an inexplicable oversight, considering the observations made by all sides that the war waged in Vietnam took place as much on mental and emotional landscapes as it did in the physical environment. In an ocean of historiography concerned with America's involvement in Southeast Asia, works concerned with America's communications initiative barely constitute a drop. Few studies even acknowledge the place of the PSYOP program in America's Vietnam War strategy, and only a couple of scholars make it the focus of their attention.[10] No account examines the JUSPAO within the context of America's historical experience of conducting PSYOP.

Studies concerned with American involvement in Vietnam rarely mention the JUSPAO and, when they do cite it, underestimate the importance of its activities. *The Vietnam War: An Almanac* provides an overview of the war's development by including a chronology of events leading up to the blood shed by all sides. It addresses various aspects of the conflict with a diverse collection of essays, and it presents biographical sketches of more than sixty of the major political and military figures from the United States, France, and Vietnam who became associated with the conflict. It cites the information program's activities with only one brief entry under June 12, 1972. It notes, "The Joint US Public Affairs Office [JUSPAO] in Saigon is closed after seven years of directing psychological warfare in Vietnam. Its duties are taken over by the USIA and other agencies; JUSPAO is declared successful in 'winning the hearts and minds' of the Vietnamese people."[11] From this account, one distinctly gets the impression that the operation's chief significance is that it closed its doors in 1972 after declaring itself a success. One is left to wonder how it earned that distinction given the ultimate demise of the RVN.

A more thorough and thoughtful exploration of the war's psychological landscape than what the *Almanac* provides is undertaken in the *Encyclopedia of the Vietnam War*, which suggests there exist "many reasons the United States and South Vietnam failed to 'win hearts and minds.'"[12] Still, contributing authors make little mention of the JUSPAO in their essays about America's pacification strategy. Dale Andrade, an American military historian of the Vietnam War, for example, defines pacification as the attempt made by the United States to extend its influence into areas "beset by insurgency."[13] America's "Other War" in Vietnam involved a massive injection of United States aid and know-how to modernize the RVN by building various public works projects such as hospitals and schools. It also included attempts to promote lasting reforms such as land redistribution and efforts to make the country side secure from guerrilla attacks. The JUSPAO does not merit one mention by Andrade, despite the agency's extensive involvement in this aspect of the war.

Andrade focuses his attention on the "singular direction" brought to pacification by Robert W. Komer, who, beginning in 1967, supervised Civilian Operations and Revolutionary Development Support (CORDS) under the authority of United States Army General William Childs Westmoreland, commander of the MACV.[14] The JUSPAO only appears in the *Encyclopedia* as part of its entry describing the United States Information Service (USIS), the overseas component of the USIA. The American social historian Paul M. Taillon notes in his brief essay that the JUSPAO "directed propaganda at the South Vietnamese people, Viet Cong guerrillas, and the North Vietnamese through radio, leaflet drops and airborne loudspeakers."[15]

While these works at least make mention of the PSYOP program, many of the military, diplomatic, and cultural histories of America's participation in Vietnam's War routinely bypass it altogether, consuming themselves instead with the strategies and tactics of conventional battles and negotiations. These studies that analyze the war, while they offer many pertinent insights into the other aspects of the way Americans conducted the conflict and conceived policy in Vietnam, tend to either fully neglect the USIA or consider its work only in passing.[16] They introduce the USIA into their discussion for one of two reasons. They cite the PSYOP either to call attention to the cultural differences that existed between Americans and Vietnamese or to scrutinize the various organizational changes that took place. Studies of the war casually describe programs the USIA carried out through the JUSPAO to support the RVN, but they rarely explore these initiatives any further. In so doing, they leave readers wondering about the precise relationship that existed between the JUSPAO program and American aspirations in Vietnam.

For example, in her synthesis *The Vietnam Wars, 1945–1990*, Marilyn B. Young evidently does not consider a single mention of JUSPAO warranted. Her discussion of the USIA is limited to less than a page of a work that totals over three hundred, and it does not chronicle agency efforts outside of those made under the John Fitzgerald Kennedy administration that occurred prior to direct American military intervention. Young indicates that the USIA sponsored a "Name the Enemy Contest" among its Vietnamese staff in 1962 and describes it as "only one of hundreds of such efforts." She acknowledges the agency's "all-important task" of building support for the RVN, but she writes little more about it.[17] Even to learn the outcome of the USIA effort to re-christen Vietnamese guerrillas, one must consult her footnotes to learn that the agency abandoned the effort after deciding the label of "Viet Cong" would suffice after all.

Young's analysis of the USIA program barely registers in her text, and she makes no mention of its later expansion under the JUSPAO in 1965. Following her brief description of a few operations, she notes that American PSYOP advisors "were often disappointed in their Vietnamese students." She characterizes the USIA endeavor as "frustrating work." As evidence, she cites an attempt allegedly made by the chief of Army Republic of Vietnam (ARVN) PSYWAR to smuggle into Vietnam five million piastres worth of pornography from Taiwan. She concludes such behavior "did not make the job any easier."[18] It is a reasonable assertion, but it does little to address the fundamental question of how Americans came to operate a program that so openly contradicted the democratic values they espoused. Nor does it attempt to gauge the relationship that existed between the operation and overall United States policy.

The lack of concern shown by scholars for the psychological program America conducted in Vietnam becomes partly understandable when one contemplates the difficulties inherent to measuring the effects of persuasive communications. As former USIA official Thomas C. Sorensen acknowledges in his examination of the agency's work, "The effect of this war of words on the enemy was never easy to determine."[19] Sorensen notes that the number of factors that can influence a person's behavior is overwhelming, and pinpointing the primary impetus of an individual's actions is necessarily imprecise, if not impossible.

A soldier might choose to defect to the other side due to any one or all of a variety of reasons. In wartime, one's action might be prompted by receiving a distressing letter from a loved one at home, experiencing a harrowing moment on the battlefield, becoming disillusioned with one's cause, disagreeing with one's superiors, reading a piece of propaganda, or simply growing weary with war. Identifying the causation of a person's political outlook, in either war or peace, is no less certain.

Americans in Vietnam did evaluate the impact made by their communications in many ways, including direct observation, counting defections to the RVN, interviewing prisoners of war, and examining captured documents to gauge the reaction of RVN opposition to individual campaigns.[20] None of these indicators, either singularly or combined, is fully satisfactory as a gauge of USIA success, because each proved dependent on a variety of subjective factors. The most widely used indicator of success in Vietnam, for example, the interrogation of prisoners through the Volunteer Informant Program, produced mixed results at best. Officials found that the feedback they received tended to be "somewhat less than accurate, less than objective, and less than completely reliable." Its usefulness varied according to the qualifications and reliability of translators, the motivations of informants, the veracity of sources, and the competency of assessors.[21]

A JUSPAO official indicated the difficulty one faces, even while present at the time campaigns are carried out, in assessing the precise impact of PSYOP. In a report from Bien Hoa, the representative queried about his agency's work:

> Is it effective? Is this concerted Chieu-Hoi [defector inducement] program what is causing the increase in the number of ralliers [defectors] or is it Allied military operations? Do the ralliers who say they have read a leaflet and/or heard a tape tell the truth or are they trying to flatter their questioner? These are the kinds of questions that the JUSPAO rep. must constantly ask. The answers are far from clear.[22]

JUSPAO director Barry Zorthian also could come up with no definitive answers regarding the value of his agency's product when questioned. He indicated in response to questions raised about his agency's effectiveness:

> It is hard to justify much of our programming throughout the world as "absolutely essential" and, if you ask me to say every piece of printing we do meets the yardstick of essentiality, I would fall flat on my face.[23]

Zorthian pointed out that gauging the effectiveness of the JUSPAO eluded quantitative measurement. It constituted a process of "constant reappraisal and adjustment" that stemmed from "thorough and considered judgments based on experience and on-the-scene requirements." In the end, Zorthian stood unwilling to support that his agency achieved "one hundred percent effectiveness out here," but he did suggest that it often proved "more discriminating than perhaps seems to be the case."[24]

Given the variety of phenomena responsible for determining individual action and the difficulties involved in weighing the impact made by each of them on people at moments of decision, the hesitancy evident among historians to analyze USIA activities in Vietnam appears reasonable. In her study of American combat propaganda in the Southwest Pacific during World War II, Allison Gilmore concedes, "The greatest difficulty in studying psychological warfare is assessing its effectiveness."[25] There exist no definitive means to quantify the level of effectiveness Americans achieved with the particular psychological programs they carried out in Vietnam. The evidence to do so conclusively does not exist. Of course, it rarely does for anything. What is telling in this case, though, is the amount of pressure to demonstrate its results that the PSYOP program constantly felt, not only from officials outside of the USIA, but as Zorthian's comments indicate, from individuals within it as well. Skepticism regarding the value of America's information program has proven historically pervasive among both the nation's public and its officials.

Isolating the reason behind a soldier's decision to defect, or discerning the influence of America's information program on a person's choice to support the RVN, are problems the passage of time only exacerbates for historians. Still, individuals who analyze the past are well accustomed to wrestling with questions they must answer using incomplete and inconclusive evidence. Other reasons exist to explain the paucity of attention scholars have provided to the PSYOP program. One plausible explanation stems from the lack of overall evidence, as, until recently, much of the material dealing with the USIA effort in Vietnam has remained classified.

Just as important, perhaps, PSYOP do not fall neatly under any single one of the categories that historians use to signify their particular fields of interest. Instead, programs like the one supervised by the USIA in Vietnam are on the periphery of studies categorized by scholars as military, diplomatic, and cultural histories. This situation is the case not only for the USIA in Vietnam, but also for the agency throughout the Cold War. The majority of works concerning the American information effort are products of former personnel who demonstrate greater interest in telling stories about their experiences than in critically analyzing their former agencies.[26]

Meanwhile, diplomatic and military historians have often treated activities in the psychological realm much like policy makers or military leaders did at the time of execution. They consider them as sort of a sidebar, one that possesses occasional anecdotal interest but offers little insight into the main problem they are considering. This situation has only begun to change recently as historians concerned with international relations have begun to identify the importance of culture in both making and carrying out policy during both war and peace.[27] As a result, a body of work is gradually developing that explores this convergence of culture and policy where it has often appeared prominent, through programs carried out by the USIA to promote international acceptance of American policies.[28]

This study posits that there is much value in setting aside the significant, but problematical, question of measuring the precise impact made by individual campaigns that the USIA carried out in Vietnam and focusing on the expectations and the ideas that shaped the overall program. Scholars have produced a wealth of material demonstrating how the political culture of the Cold War influenced American film, literature, religion, television, social relationships, and even comic books.[29] Yet they have made little effort to explore the link between American Cold War culture and the methods employed overseas by the United States to communicate its ideals and its policies with others. The decided focus of those individuals who have examined America's general Cold War information program has remained on its effectiveness in undermining communism, despite the hazards contained in this task.[30]

These works are insightful, even without fully demonstrating whatever case they make about the exact impact of American psychological initiatives. In particular, they provide institutional histories that illustrate the relationship between the USIA and American foreign policy. Employing such a myopic focus, however, wards off questions regarding the fundamental assumptions Americans held that buttressed these programs. It inhibits understanding what officials hoped their campaigns would achieve and neglects to identify what role these activities played in shaping American perceptions. Such an approach fails to consider the information program's subculture, the distinct set of values and beliefs shared among its members.[31] It is an area of exploration that has been fruitfully canvassed by scholars regarding other groups and organizations that fulfill America's foreign policy process.[32]

Nancy E. Bernhard utilizes this approach in her uniquely perceptive analysis of the United States State Department's Domestic Information Campaign. She avoids the inclination to become preoccupied with the program's effectiveness and opens avenues that reveal an area where diplomatic history and cultural history intersect. In her essay, Bernhard demonstrates how anticommunism pervaded American political culture to the point where the contradiction inherent in employing tactics of distortion in the defense of democratic values eluded policy makers during the early years of the Cold War. Even as they condemned the propaganda emanating from the communist world, American officials utilized strategies designed to preempt any open discussion at home about America's Cold War policies.[33]

Concentrating on the values that shaped the USIA involvement in Vietnam builds on Bernhard's pioneering work. Analyzing America's information program to discern how its officials viewed their nation's role in the conflict it had entered, how they intended to utilize the USIA to advance their ideas, and how their program evolved from the ways they imagined American and Vietnamese relations yields important insights. Most prominently, such scrutiny reveals that the ideas manifest in the information program both reflected and further contributed to the confusing, difficult, and ultimately tragic experience of America's intervention in Vietnam. It makes clear that in Vietnam American officials conflated their rhetoric and the reality that confronted them in a way that prevented them from seriously reassessing their fundamental assumptions. In the process, they deemed it their exclusive right to utilize undemocratic means to secure an end that they had identified as a democratic purpose.

MAKING BAD POLICY PALATABLE

The timing is right to revisit the operation that the USIA carried out through the JUSPAO in Vietnam. Many of the documents conveying what Americans hoped to accomplish with this agency have only recently become available through the declassification process. Analyzing this material yields fresh insights not only regarding America's intervention in Vietnam, but also relating to the ongoing effort of the United States to communicate its intentions in a world grown skeptical of its ambitions. While historical parallels are always elusive and imprecise, the end of the Cold War has ushered in a time when, similar to the defensiveness it exhibited in Vietnam, the United States finds itself straining to persuade others of the integrity of its policies and of the goodwill behind its actions.

Americans remain as committed to the idea of "nation building" in the present as they were in the past and, at least in theory, PSYOP comprise a key component of this policy. Yet, many of the same questions that commentators raised about the United States effort in Vietnam are once again being asked by observers of American endeavors in Iraq and Afghanistan. The primary criticisms leveled at the program in Vietnam by postwar government sponsored studies—that the JUSPAO remained understaffed and poorly funded, and that it emphasized mass communication techniques at the expense of face to face communications and the promotion of policies responsive to popular concerns—remain evident regarding American operations abroad today.[34] In a recent article appearing in *Time* magazine, one Pentagon official labels American PSYOP as "a broken force," and a battalion commander calls his unit "criminally underequipped." The article makes clear that the current initiative, carried out mainly by reservists who lack language training, hinges on the construction of "leaflets and pro-U.S. broadcasts aimed at winning the 'hearts and minds' of citizens."[35]

Even more telling is a candid admission made recently by senior American commanders in Iraq, which suggests the United States took little heed of its experience with PSYOP in Vietnam. Then United States Army Major General Peter W. Chiarelli and Major Patrick R. Michaelis note that "the traditional military training model" prepared them to win America's "wars on the plains of Europe or the deserts of the Middle East." It enabled them to envision "large, sweeping formations; coordinating and synchronizing the battlefield functions to create that 'point of penetration'" for rapid exploitation. These commanders have recognized, though, that "in Baghdad, that envisioned 3-decade-old concept of reality was replaced by a far greater sense of purpose."

That is, "Synchronization and coordination of the battlespace was not to win the war, but to win the peace." For American commanders, penetration has come to mean "local infrastructure improvement; training of security forces, understanding and educating the fundamentals of democracy; creating long lasting jobs that would carry beyond short-term infrastructure improvement."[36] This one is a valuable lesson, but it is also one that has been belatedly learned. By apparently neglecting the experience in Vietnam, American political leaders and commanders have come to recognize the value of PSYOP only after they have spent much of their nation's political capital, both at home and abroad. It remains to be seen whether Osama Sibliani, the publisher of *Arab American News*, has correctly identified the full consequences of this failure. He contends that "The United States could have the Prophet Muhammad doing public relations and it wouldn't help."[37]

This examination of the notions that guided United States communications in Vietnam is a case study that makes clear for Americans the hazards inherent to dismissing dialogue about national policy preemptively for the sake of such things as political expediency and self-assurance. Through the USIA program, American leaders had at their disposal in Vietnam the means to create a more sensitive policy than the one that they executed. On paper, PSYOP provided them with the capacity to detect and to respond to popular Vietnamese concerns, regarding both RVN rule and the endorsement provided it by the United States. In reality, American officials possessed no tradition of using a PSYOP program in this manner and proved unable to reshape the political culture they inherited.

If it is measured by the standards set for it by American officials at the time of its operation, the USIA-directed program proved ineffectual in Vietnam. It failed to foster enough support within either Vietnam or the international community to prevent the downfall of the government in Saigon that America favored. The few who have commented on the USIA have suggested its PSYOP had little chance for overall success because of what they deem the bankruptcy of American policy in Vietnam. The refrain goes, "Propaganda cannot make bad policy palatable."[38] While acknowledging this point, that PSYOP personnel must play the cards they are dealt, so to speak, this study responds by advancing the argument that America's PSYOP program played a larger role than existing analysis appreciates in perpetuating the American mindset that led to the dubious initiative in the first place. The impact made by particular PSYOP campaigns remains debatable, but it is undeniable that the overall American program did little to encourage the multidirectional dialogue about the war that United States leaders claimed the JUSPAO would facilitate.

American officials had created the JUSPAO as a separate policy-making entity that complemented efforts in military, economic, and political matters. It became one of the major components of the United States Mission, which included the Embassy for political action, USMACV for military affairs, and USAID for economic policy. The JUSPAO carried out operational and coordination activities until the 1967 establishment of CORDS under General Westmoreland, which assumed responsibility over the entire pacification program that Americans carried out in conjunction with the RVN to gain the support of Vietnamese civilians for Saigon's leadership. Under the new arrangement, the JUSPAO provided policy guidance to the Psychological Operations Division of CORDS (CORDS/POD).[39]

Throughout the war, the JUSPAO also provided guidance through the USMACV, Psychological Operations Directorate to all United States Army and other military PSYOP elements. The army's effort involved the 4th PSYOP Group (Saigon), which directed the following battalions: the 6th in Bien Hoa, the 7th in Da Nang, the 8th in Nha Trang, and the 10th in Can Tho. These units provided tactical support to field force commanders, encouraged defection and surrender in both the NLF and the People's Army of Vietnam (PAVN), and informed Vietnamese generally about RVN programs.[40] The JUSPAO also advised the United States Naval Forces, Vietnam (NAVFORV), which combined PSYOP with its routine offshore and inland waterway operations, and it offered guidance to the 7th United States Air Force, which dropped leaflets and broadcast messages from the air.[41]

With its wide-ranging authority over the operational components, both military and civilian, the JUSPAO ostensibly possessed great influence in the development and execution of American PSYOP. As an element of the United States Mission, it also sat located to influence United States policy in general, offering its advice regarding the potential psychological implications of discussed policies. "Under the heading of staff functions," former agency director Zorthian writes, JUSPAO responsibilities involved the need "to identify and determine policies." He calls this agency function a "bold and active role" and argues of its importance:

> For while the popular appeal of particular actions cannot always be the controlling factor in the adoption of policy, it should certainly constitute a major consideration and should be ignored only for the most overriding reasons. Further, if the factor of appeal must be overlooked, the communicator can often offer modification or method of presentation that will increase its acceptability.[42]

Zorthian contends that in Vietnam his agency's "participation in the American effort came closer to the desirable than has usually been the case." Still, he acknowledges that its impact "varied with different direction in the Mission." The JUSPAO managed to attain its most consistent influence by assuming the role of "honest broker." It acted to check the worst impulses of

America's military by "monitoring—policing if you will" America's compliance with the ground rules for military operations. Zorthian believes this function proved "essential," and he notes that since the rules "were designed to have the most positive effect possible on the population, this honest broker's role was important."[43]

While Zorthian's observation regarding the importance of the "honest broker's role" is an astute one, it is clear that the influence exerted by JUSPAO on the policy making process represented a minimal contribution at best. Historically, neither America's policy makers nor its military leaders have proven eager to incorporate PSYOP into their activities, and the nation's experience in Vietnam fails to stand out as an exception. Among United States military leaders, PSYOP have long been "narrowly defined in terms of leaflets and broadcasts" and considered a "minor appendage" to conventional forces.[44]

In preparing for a potential invasion of Japan in 1944, for example, the United States Joint Chiefs of Staff (JCS) responded to a proposed psychological initiative in a typically negative fashion. "At a time when we are unable to plan military operations for want of service troops, it hardly seems justifiable to divert such to a project of this nature."[45] While the military did utilize PSYOP in Vietnam, its leaders viewed it as an occasional supplement to traditional tactics and strategy. Military leaders regarded guidance received from the JUSPAO lightly, as they held "anyone, themselves included, could write leaflets."[46] Even by the 1970s, almost two decades after the first American military advisors arrived in Vietnam, United States Army manuals remained "almost completely silent on the role of an individual soldier as a PSYOP participant."[47]

The relationship maintained between America's information officials and its diplomats is categorized by at least as much acrimony. During both world wars, and throughout the Cold War, information officials found their agencies accused by the United States Department of State of improperly meddling in policy making and locked out of the process all together.[48] If anything, the situation worsened in Vietnam under a presidential administration that conceived of the information program primarily as a means to secure better public relations with its domestic constituency by first and foremost utilizing the JUSPAO as a forum to publicize America's "Other War."

The strategy President Johnson and his advisors outlined for the JUSPAO amounted to deflecting attention from American setbacks and casualties in the military conflict by publicizing the "Other War" in hopes of providing both a sense of accomplishment and a cause for optimism. At a meeting with his cabinet in the summer of 1965, Johnson asked, "How can we get everybody to compete with [Secretary of Defense] McNamara in the press?" He cited the need to "constantly remind the people that we are doing other things besides bombing." Leonard Unger, deputy assistant secretary of state for Far Eastern affairs and head of the Vietnam Task Force, informed Johnson he had already taken this question up with Zorthian and his press people.[49]

Meanwhile, in Vietnam, the JUSPAO often did not even receive direction about what guidance already existed, let alone shape policy. Reuben S. Nathan, JUSPAO Chief of Planning (1965–1966), asserts that he negotiated his entire two tours of duty with no more than "a single piece of guidance." He contends further that this minimal amount of direction came to him "orally by a high ranking State Department official," and it called on him to "de-emphasize the American presence in Vietnam." He not only labels the policy as "unworkable," but also notes also that, in light of the obvious American military escalation under way in 1965, it proved "so counterproductive that I was compelled to disregard it."[50]

United States policy makers, despite the apparent mandate they gave the PSYOP program and the rhetorical support they offered it to play a role in planning policy, proved mainly interested in using their communications apparatus to engage in a monologue that acted to stymie reflective and flexible decision-making. Regardless of the stories told by former PSYOP officials relating to the success of particular propaganda campaigns, the program they carried out in Vietnam did not deliver on its enormous potential as a foreign policy tool. American officials at the time hailed the JUSPAO operation as a radical departure for America's international communications, but it came to serve largely as a mouthpiece for official views that managed to make little of its intended impact on audiences.

Through the course of its existence, the JUSPAO became a propaganda agency, the kind of thing American officials profess to abhor, but nonetheless the sort of tool they have repeatedly resorted to using throughout their history. This process occurred largely because of the very notion officials used to rationalize their use of PSYOP. By the time they arrived in Vietnam, American leaders had convinced themselves they possessed a uniquely objective viewpoint on world affairs. They believed that their PSYOP program differed from those carried out by other nations. They perceived it as deriving only from the facts of a matter they presumed themselves alone in being able to discern for others.

The idea of American objectivity became so ingrained in the USIA program that officials lost sight of its subjective nature and used it as an unabashed proponent of policy. In the process, it became fully un-American, a voice that projected its righteous tone so loudly that it became incapable of considering views expressed by others. It manifested for the world to witness the antithesis of the freedoms Americans claimed to be fighting for in Vietnam, and it bolstered the case of critics who cited the gap between the rhetoric and the reality of United States policy. The *New York Times* directly raised the question on its editorial page by asking whether the American posture constituted "Truth or Propaganda?" The newspaper rendered its judgment that the USIA engaged in the latter and opined that, as a result, "the credibility of the United States Government has been one of the numerous casualties of the war in Vietnam."[51]

Playing the "What if?" game in history is hampered by unseen variables that obscure one from concretely discerning the endpoints of paths not taken. It is not the point of this study to suggest that if American officials had conducted PSYOP in Vietnam differently they would have secured a more successful conclusion to the war than what they actually managed to bring about. It does make clear, though, that the USIA program carried out by the JUSPAO in Vietnam did not deliver on its potential to inform American policy by thoughtfully canvassing the Vietnamese psychological landscape.

United States officials, in Washington and Saigon, too often stood content with the ahistorical portrait of Vietnam painted for them by its former French colonizers and the anthropologists that the JUSPAO relied on for advice regarding the character of Vietnamese.[52]

Americans too readily accepted the idea that Vietnam represented a blank slate and demonstrated greater concern with molding Vietnamese opinion than with soliciting it.[53] While an alternative outcome would have been far from a certainty, it is evident that had Americans utilized the USIA to gather information about the Vietnamese and their concerns instead of simply to disseminate materials, other possibilities existed; at the very least, Americans would have possessed a more sensitive barometer than any they actually employed to gauge the climate of opinion in Vietnam.

NOTES

1. Report, Psychological Operations, "Conduct of the War: Functional Analysis, Volume VI, Book 2," May 2, 1980, BDM Corporation Study: A Study of Strategic Lessons, box 4, LBJL.
2. Zorthian, "The Use of Psychological Operations," 40.
3. Zorthian, "The Use of Psychological Operations," 40.
4. Summary Record of Meetings, Honolulu, June 2, 1964, *FRUS, 1964–68, Vietnam 1964*, I: 429.
5. Article, "Honor Awards Presented in Saigon," *USIA Correspondent*, March 1966, "USIA Correspondent 1965–1967," box 14, Papers of Leonard Marks, LBJL.
6. Article, "Honor Awards."
7. Article, "Honor Awards."

8. Congressperson David S. King, *United States of America Congressional Record: Proceedings and Debates of the 89th Congress, Second Session* (Washington, DC: Government Printing Office, 1966), Volume 112—Part 5, 5874.

9. Congressperson King, *United States of America Congressional Record*, 5874.

10. Chandler, *War of Ideas*; Latimer, *U.S. Psychological Operations in Vietnam*.

11. John S. Bowman, ed., *The Vietnam War: An Almanac* (New York: World Almanac Publications, 1985), 315.

12. Ngo Vinh Long, "Vietnamese Perspectives," in Stanley Kutler, ed., *Encyclopedia of the Vietnam War* (New York: Scribner's, 1996), 592.

13. Dale Andrade, "Pacification," in *Encyclopedia of the Vietnam War*, 417.

14. The concern Andrade demonstrates for the organizational aspects of pacification is also a hallmark of other studies devoted to America's "Other War," most notably Richard A. Hunt, *Pacification: The American Struggle for Vietnam's Hearts and Minds* (Boulder, San Francisco, and Oxford: Westview Press, 1995); Robert W. Komer, *Bureaucracy Does Its Thing: Institutional Constraints on U.S.-G.V.N. Performance in Vietnam* (Santa Monica, California: Rand Corporation, 1972).

15. Paul M. Taillon, "United States Information Service," in *Encyclopedia of the Vietnam War*, 554.

16. Examples include Loren Baritz, *Backfire: A History of How American Culture Led Us into Vietnam and Made Us Fight the Way We Did* (New York: Ballantine Books, 1985); Eric M. Bergerud, *The Dynamics of Defeat:The Vietnam War in Hau Nghia Province* (Boulder: Westview Press, 1991); Douglas Blaufarb, *The Counterinsurgency Era* (New York: Free Press, 1977); Larry E. Cable, *Conflict of Myths: The Development of American Counterinsurgency Doctrine and the Vietnam War* (New York: New York University Press, 1986); Chester Cooper, *The Lost Crusade: America in Vietnam* (New York: Dodd, Mead, 1970); Phillip B. Davidson, *Vietnam At War: The History, 1946–1975* (New York and Oxford: Oxford University Press, 1991); Frances Fitzgerald, *Fire in the Lake: The Vietnamese and the Americans in Vietnam* (Boston: Little, Brown & Co., 1972); George C. Herring, *LBJ and Vietnam: A Different Kind of War* (Austin: University of Texas Press, 1994); Kahin, *Intervention*; Stanley Karnow, *Vietnam: A History* (New York: The Viking Press, 1983); Gabriel Kolko, *Anatomy of a Wa:, Vietnam, the United States and the Modern Historical Experience* (New York: Pantheon Books, 1986); Andrew F. Krepinevich, *The Army and Vietnam* (Baltimore, Maryland: The Johns Hopkins University Press, 1986); Michael E. Peterson, *The Combined Action Platoons: The U.S. Marines' Other War in Vietnam* (Westport, Connecticut: Praeger, 1989), Jeffrey Race, *War Comes to Long An* (Berkeley and Los Angeles: University of California Press, 1973); D. Michael Shafer, *Deadly Paradigms: The Failure of U.S. Counterinsurgency Policy* (Princeton, New Jersey: Princeton University Press, 1988), Harry Summers Jr., *On Strategy: A Critical Analysis of the Vietnam War* (Novato, California: Presidio Press, 1982); Thomas C. Thayer, *War Without Fronts* (Boulder: Westview Press, 1985); and Marilyn B. Young, *The Vietnam Wars, 1945–1990* (New York: HarperPerennial: A Division of HarperCollinsPublishers, 1991).

17. Young, *The Vietnam Wars*, 92–93.

18. Young, *The Vietnam Wars*, 92–93.

19. Sorensen, *The Word War*, 288.

20. Taro Katagiri, "A Former PSYOP Group Commander in Vietnam Looks Back," in *The Art and Science of Psychological Operations*, I: 137–44.

21. R. P. Morris, "Psychological Operations in Vietnam: Indicators of Effectiveness at the U.S. Army Division Level," in *The Art and Science of Psychological Operations*, II: 946–57.

22. Report, II FFORCEV/BIEN HOA, undated, "Reports of JUSPAO Field Representatives for the month ending August 31, 1966," September 20, 1966, Counterinsurgency File (CI), Richard M. Lee Papers, MHI.

23. Memo, to USIA director Leonard H. Marks from JUSPAO director Barry Zorthian, November 12, 1966, "Material on JUSPAO," box 4A39, Papers of Barry Zorthian, LBJL.

24. Memo, to USIA director Leonard H. Marks from JUSPAO director Barry Zorthian, November 12, 1966.

25. Allison B. Gilmore, *You Can't Fight Tanks with Bayonets: Psychological Warfare against the Japanese Army in the Southwest Pacific* (Lincoln and London: University of Nebraska Press, 1998), 146.

26. See Edward W. Barrett, *Truth Is Our Weapon* (New York: Funk and Wagnalls, 1953); Leo Bogart, *Premises for Propaganda: The United States Information Agency's Operating Assumptions in the Cold War* (New York: The Free Press, 1976); Robert Elder, *The Information Machine: The United States Information Agency and American Foreign Policy* (New York: Syracuse University Press, 1968); Fitzhugh Green, *American Propaganda Abroad* (New York: Hippocrene Books, 1988); John W. Henderson, *The United States Information Agency* (New York: Frederick A. Praeger Publishers, 1969); Michael Nelson, *War of the Black Heavens: The Battles of Western Broadcasting in the Cold War* (Syracuse: Syracuse University Press, 1997); Alvin Snyder, *Warriors of Disinformation: American Propaganda, Soviet Lies, and the Winning of the Cold War, an Insiders Account* (New York: Arcade Publishing, 1995); Sorensen, *The Word War*; Hans N. Tuch, *Communicating with the World* (New York: St. Martin's Press, 1990); Randolph Wieck, *Ignorance Abroad: American Educational and Cultural Foreign Policy* (Westport, Connecticut: Praeger Publishers, 1992).

27. Baritz, *Backfire*; Bradley, *Imagining Vietnam*; Jongsuk Chay, ed., *Culture and International Relations* (New York, Westport, Connecticut, and London: Praeger Publishers, 1990); Frank Costigliola, *Awkward Dominion: American Political, Economic, and Cultural Relations with Europe, 1919–1933* (New York: Hill and Wang, 1982); John W. Dower, *War Without Mercy: Race and Power in the Pacific War* (New York: Pantheon Books, 1986); Robert Fyne, *The Hollywood Propaganda of World War II* (Metuchen, New Jersey: The Scarecrow Press, Inc., 1994); Michael Hunt, *Ideology and U.S. Foreign Policy* (New Haven: Yale University Press, 1987); Amy Kaplan and Donald Pease, eds., *Cultures of United States Imperialism* (Durham, North Carolina: Duke University Press, 1993); Anne McClintock, *Imperial Leather: Race, Gender, and Sexuality in the Colonial Context* (New York: Routledge, 1995); Emily S. Rosenberg, *Spreading the American Dream: American Economic and Cultural Expansion, 1890–1945*, (New York: Hill and Wang, 1982); Edward Said, *Orientalism* (New York: Vintage, 1978).

28. See Laurien Alexandre, *The Voice of America: From Détente to the Reagan Doctrine* (Norwood, New Jersey: Ablex Publishing Corporation, 1988); Nancy E. Bernhard, "Clearer than Truth: Public Affairs Television and the State Department's Domestic Information Campaigns, 1947–1952," *Diplomatic History* 21, no. 4 (fall 1997): 545–67; Walter L. Hixson, *Parting the Curtain: Propaganda, Culture, and the Cold War, 1945–1961* (New York: St. Martin's Griffin, 1997, 1998); Richard Kuisel, *Seducing the French: The Dilemma of Americanization* (Berkeley, Los Angeles, and London: University of California Press, 1993); Reinhold Wagnleitner, *Coca-Colonization and the Cold War: The Cultural Mission of the United States in Austria after the Second World War* (Chapel Hill and London: The University of North Carolina Press, 1994).

29. See Paul Boyer, *By the Bomb's Early Light: American Thought and Culture at the Dawn of the Atomic Age* (New York: Pantheon, 1995); Elaine Tyler May, *Homeward Bound: American Families in the Cold War Era* (New York: BasicBooks, 1988); Lary May, ed., *Recasting America: Culture and Politics in the Age of Cold War* (Chicago and London: The University of Chicago Press, 1989); Alan Nadel, *Containment Culture: American Narratives, Postmodernism, and the Atomic Age* (Durham: Duke University Press, 1995); William W. Savage, *Comic Books and America, 1945–1954* (Norman and London: University of Oklahoma Press, 1990); Stephen J. Whitfield, *The Culture of the Cold War* (Baltimore and London: The Johns Hopkins University Press, 1991).

30. See Hixson, *Parting the Curtain*; Nelson, *War of the Black Heavens*; Sig Mickelson, *America's Other Voice: The Story of Radio Free Europe and Radio Liberty* (New York: Praeger, 1983); Arch Puddington, *Broadcasting Freedom: The Cold War Triumph of Radio Free Europe and Radio Liberty* (Lexington: The University Press of Kentucky, 2000), Alvin A. Snyder, *Warriors of Disinformation*; Sorensen, *The Word War*; Wieck, *Ignorance Abroad*.

31. John P. Lovell, "The United States as Ally and Adversary in East Asia: Reflections on Culture and Foreign Policy," in *Culture and International Relations*, 91.

32. Burton Hersh, *The American Elite and the Origins of the CIA* (New York: Charles Scribner's Sons, 1992); Morris Janowitz, *The Professional Soldier: A Social and Political Portrait* (New York: The Free Press, 1960); E. J. Kahn Jr., *The China Hands: America's Foreign Service Officers and What Befell Them* (New York: Penguin Books, 1976); Peter Karsten, *The Naval Aristocracy: The Golden Age of Annapolis and the Emergence of Modern American Navalism* (New York: The Free Press, 1972); Andrew F. Krepinevich Jr., *The Army in Vietnam*; Charles C. Moskos Jr., *The American Enlisted Man* (New York: Russell Sage, 1970); Martin Weil, *A Pretty Good Club: The Founding Fathers of the U.S. Foreign Service* (New York: Norton, 1978).

33. Nancy E. Bernhard, "Clearer than Truth: Public Affairs Television and the State Department's Domestic Information Campaigns, 1947–1952," *Diplomatic History* 21, no. 4 (fall 1997): 545–67.

34. Report, Psychological Operations, "Conduct of the War: Functional Analysis," Volume VI, Book 2.

35. "Still Short in Iraq," March 14, 2005, *Time* magazine, 13.

36. U.S. Army Major General Peter W. Chiarelli and Major Patrick R. Michaelis, "Winning the Peace: The Requirement for Full Spectrum Operations," *Military Review* (July–August 2005): 4.

37. United States Army Lieutenant Colonel Steven Collins, "Mind Games," *PSYWARRIOR*, http://www.psywarrior.com/MindGamesSCollins.html (May 19, 2007).

38. Sorensen, *The Word War*, 292.

39. Report, Psychological Operations, "Conduct of the War."

40. Taro Katagiri, "The 4th PSYOP Group: Organization, Operations, and Observations," in *The Art and Science of Psychological Operations*, I: 220.

41. Ronald De McLaurin, Carl F. Rosenthal, Sarah A. Skillings and others, "PSYOP in Vietnam: A Many Splintered Thing," in *The Art and Science of Psychological Operations*, I: 225.

42. Paper delivered at the National Strategy Information Center Conference by Barry Zorthian, The Use of Psychological Operations in Combatting "Wars of National Liberation," March 11–14, 1971, "Material on JUSPAO," Papers of Barry Zorthian, 4A39, LBJL.

43. Paper delivered by Barry Zorthian, "Wars of National Liberation."

44. Alexander Askenasy, "Role of Psychological Operations Within the Military Mission," in *The Art and Science of Psychological Operations*, I: 67.

45. Operational Plans for OSS Activities in Pacific Ocean Areas, Memorandum by the Commander in Chief, United States Fleet and Chief of Naval Operations, December 14, 1944, *Records of the JCS, Part I: 1942–1945, The Pacific Theater*, viii, F267.

46. Report, Psychological Operations, 13–27.

47. Francis M. Watson Jr., "PSYOP and Related Activities," in *The Art and Science of Psychological Operations*, I: 58.

48. See Henderson, *The United States Information Agency*; Hixson, *Parting the Curtain*; Holly Cowan Shulman, *The Voice of America: Propaganda and Democracy, 1941–1945* (Madison: University of Wisconsin Press, 1990); Sorensen, *The Word War*; and James P. Warburg, "The Moronic Little King Incident," in *A Psychological Warfare Casebook*, ed. William E. Daugherty (Baltimore: Johns Hopkins Press, 1958).

49. Notes, Cabinet Room, July 21, 1965, "July 21–27, 1965 Meetings on Vietnam," box 1, Meetings Notes File, LBJL.

50. Reuben S. Nathan, "Making Policy is Not the Propagandist's Business—Or Is It?" in *The Art and Science of Psychological Operations*, I: 125.

51. "Truth or Propaganda?" *New York Times*, April 23, 1965, 34.

52. Bradley, *Imagining Vietnam*.

53. See Jefferson P. Marquis, "The Other Warriors: American Social Science and Nation Building in Vietnam," *Diplomatic History* 24, no. 1 (winter 2000): 79–105.

Chapter Three

The Great Cultural Divide

OVERVIEW

This chapter is divided into three parts. The first, entitled "Over Five Billion Served," analyzes the memoirs of those individuals who carried out the PSYOP program in Vietnam. It proposes that, while these works provide great insight into the organization and the daily operations of both the USIA and the JUSPAO, they provide little scrutiny of the message these agencies delivered to Vietnamese. The chapter's second section, "Lassitude Will Be Lassitude," builds on the previous discussion by suggesting much can be learned about the assumptions that provided the foundation for the American PSYOP program in Vietnam by closely considering how its planners have since rationalized its inability to muster sufficient Vietnamese support for the RVN.

In "A Cultural Box," this chapter's third and final section, the allegation made by some scholars that Americans ignored Vietnamese culture is analyzed by looking at the assessments of Vietnamese culture rendered at the time of the war by American officials. In sum, this chapter asserts that the differences that Americans perceived between themselves and Vietnamese sprang more from American imaginations than from a Vietnamese historical process they proved unwilling to consider, both during and after the war.

OVER FIVE BILLION SERVED

There is much to learn from the manner by which the ideological and institutional settings provided by the Cold War shaped the American PSYOP program in Vietnam. The USIA endeavor resulted from the expectations Americans had formed regarding the uses of information and propaganda, and from the context they provided to the war. Americans took with them to Vietnam preconceived ideas about how best to employ PSYOP. Focusing on these assumptions and intentions Americans held for their PSYOP program is an instructive departure from the few works that do consider the USIA effort in Vietnam in more than a cursory way.

Written by former agency officials, these accounts typically contain the reflections of their authors, who comment about their Vietnam experience in the context of their careers in general.[1] They provide an insider's perspective on the part the USIA played during the war and yield an invaluable glimpse into the agency's daily operations. They are useful for enabling their readers to establish a vivid connection with the frustration their authors felt working with people they considered so different from themselves. They also provide an indication of the difficulties encountered by information officials in interacting with American and Vietnamese colleagues, whom they allege held a low opinion of their agency's worth to the war effort.

The source of the strength of these studies offered by former officials is also the root cause of their weakness; they are unable to escape the restraints placed on them by their authors' participation in the experience they describe. These works provide only a vague overview of the information program before narrowing to concentrate on the individual part played in various operations by their authors. They express little interest in scrutinizing the assumptions behind the ideas Americans employed in Vietnam, and they ultimately fail to substantiate the opinions they do manage to render for their readers about the program's effectiveness, as they consistently rely on anecdotal evidence for support.

Memoir accounts tend to deliver ambivalent judgments regarding the final success of America's psychological program in Vietnam. They offer up specific instances of achievement, usually campaigns their authors participated in carrying out, but generally they fail to consider the overall program's fundamental approach. While they consistently laud the organizational characteristics of the program, they identify uncooperative Vietnamese, skeptical American military personnel, and an ambiguous pledge of support from United States officials as obstacles the USIA simply could not overcome. In some ways these examples are revealing. Most significantly, they permit one to see how those personnel actually involved in the information campaign came to perceive retrospectively what and who acted to impede their mission. They also provide an indication of what programs the USIA emphasized and which ones those who carried them out considered most important to fulfilling America's Vietnam agenda.

In these author's haste to attribute blame to others for what they consider as America's failure to win Vietnamese support for the RVN, the proverbial forest consistently eludes their memoir accounts. Their focus remains fixed only on individual trees. In this respect little changed for these former officials between the time they spent in Vietnam and the point when they chose to write about their experiences. In looking back on their tenure with the USIA, they continue to refrain from examining the ideas that guided their efforts. They reference the rhetoric that framed their operations, and they call attention to the diligence and good intentions exhibited by information officials. They demonstrate the presumable impact of America's program by describing the size of its operations and by citing the sheer volume of material it managed to disseminate throughout Vietnam's countryside.

Typically, the memoir accounts acknowledge the difficulties contained in the task of evaluating the operations they conducted. Consistently, though, they boil the program down to its numbers, viewing its impact as somehow proportional to its size. Former JUSPAO Chief of Plans, Policy and Research Harry Latimer is one of the few to recognize the fallacy of mistaking quantity for quality. He writes, "In the absence of any real knowledge and criteria of effectiveness, we tried everything, in unlimited quantity. Witness the 5 billion leaflets in one year."[2]

By concentrating on the American PSYOP program's organizational structure and the quantity of its output, the memoir-oriented studies neglect to consider questions about what Americans had to say, how they decided to say it, and, most significantly, how what they said matched up against the actions they took elsewhere. Those who comment about the PSYOP program assume the message Americans took with them to Vietnam possessed the universal resonance those who crafted it believed it had, and they never question whether their nation's rhetoric remained in concert with the actions it took in support of the RVN. These accounts are preoccupied by questions they raise regarding the will exhibited by Vietnamese to rid their nation of communists and about the ability of Vietnamese to respond positively to enlightened American guidance. As many Americans did during the war, they also narrowly depict PSYOP as consisting of leaflets, broadcasts, campaign slogans, and civic works that somehow remained divorced from larger issues weighing on post-colonial Vietnamese hearts and minds such as national unification, independence, and socioeconomic reform through the redistribution of land.[3] To do otherwise, of course, would suggest the existence of Vietnamese discontent in the south with the governments America supported in Saigon and question the very reason United States officials used to justify being in Vietnam in the first place: to check against aggression from communists in Hanoi.

Robert Chandler's *War of Ideas* stands out among the memoir accounts, as it offers an unusually sophisticated appraisal of America's information program in Vietnam. It is at once the most refined and the least narrow treatment of America's PSYOP yet produced. Chandler, a political scientist specializing in national security affairs and international political communications, served in Vietnam as an intelligence officer in the United States Air Force. His work draws largely from USIA-generated materials instead of relying on memories and anecdotes; it thoughtfully weighs the cultural disparities between Americans and Vietnamese, and it carefully analyzes the themes employed by the United States to make its case for RVN rule.

While Chandler does spend an inordinate amount of time considering the administrative structure of the USIA, unlike most others during the war and since it ended, he moves beyond organizational, technical, and material components of the psychological program to consider what Americans had to say. He supplements his text with over ninety-five illustrations of JUSPAO leaflets and other printed materials. He proposes Americans never had a chance to engage Vietnamese people successfully in any type of meaningful communications process. Because of their inescapable status as outsiders, and because of the profound cultural disparity existent between Vietnamese and Americans, Chandler argues United States efforts could inevitably produce only negligible results.[4]

Chandler's reach is commendable, as it spans the totality of the information program Americans established to communicate with Vietnamese on both sides of the conflict. His study describes the program's organization, examines its operation, and places its discussion squarely in the context of the war. It does not skip from incident to incident, dwelling on either individual successes or failures; instead, his work evaluates the whole operation's effectiveness in supporting American policy. Most important, Chandler's analysis stresses the important role the war of ideas played in the Vietnam conflict. The author succeeds in making lucid a notion too often missing from accounts of the war. While victory ultimately went to the side prevailing in the last military battles, "the crucial factor was which side would convince the Vietnamese that its cause was just, it was the legitimate successor to the nation's pre-colonial legacy, and it could best fulfill their needs." Chandler succinctly captures what too often eludes authors who pen accounts that explore the Vietnamese conflict: "The people ultimately determined the victor; mass support brought with it the power needed to win the military war."[5] He clearly illustrates the armed struggle in Vietnam directly derived from the contesting visions Vietnamese held in their minds regarding their nation's future.

Chandler's effort to call attention to America's mostly neglected war of ideas in Vietnam alone makes his work a valuable contribution to the literature. Still, it yields little insight into the fundamental assumptions Americans held about the program they created. Chandler provides little explanation of how Americans who participated in it considered the role of the information campaign in relation to their nation's overall effort. Instead, because the sources he needed to do anything more remained classified when he published his work, he confines his analysis to describing its operation and output, and to relying on field reports to weigh in on effectiveness. He omits discussion relating to the part American information officers played in handling the media in Vietnam.

Further, Chandler provides no glimpse into the reasoning of American leaders for conducting the operation and leaves his readers to wonder precisely where the impetus for the program came from or how individuals in Washington envisioned it would operate. As a political scientist, and not as a historian, he is most concerned with how the program worked in Vietnam. He demonstrates little interest in identifying the operation's historical antecedents. There is no exploration of the thinking that spawned the program, only descriptions of what it produced and speculation regarding its effectiveness. Beyond a general overview of America's historical use of propaganda, he fails to place America's Vietnam operation within the development of United States Cold War information history. His work offers no ideas about whether Americans in Vietnam adhered to or departed from the model that the USIA provided them, or if their ideas proved continuous with the evolution of American perceptions regarding government information programs following World War II.

Chandler's thesis initially appears clear and decisive. It asserts that America's "participation in the Vietnamese war of ideas was doomed from the start." Chandler makes a case that America's intervention stood no chance of success from the beginning because the United States faced the insurmountable obstacle of overcoming its status as a "foreign invader" in the eyes of the Vietnamese. "As foreigners," he observes, "Americans could not produce national unity for the Vietnamese." America's informational campaigns failed to achieve their intended results because "it was precisely such an anti-Communist nationalism that the Americans tried to create when they interjected themselves into the Vietnamese civil war."[6] According to Chandler, only Vietnamese initiative could have made a positive difference in Vietnam, and culture rendered such a contribution impossible. He observes that Vietnamese "passivity is deeply imbedded in the nation's social and political fabric."[7] He believes Americans made their greatest mistake in failing to acknowledge they had entered an impossible situation. They had engaged in "dangerous and possibly counter productive activities by participating directly in the Vietnamese battle for Vietnamese minds."[8]

Reasons to question Chandler's case stem from the very evidence he uses as the foundation for his work. A close inspection of Chandler's footnotes illustrates that he makes an argument he cannot sustain with the material he utilizes to advance it. His sources are entirely American generated, but he insists on using them to understand how the Vietnamese thought about America's attempts to reach them. Most glaringly, he constructs a worldview of Vietnamese peasants by using such material as *The Vietnamese Peasant: His Value System* and *Americans and Vietnamese: A Comparison of Values in Two Cultures*. The first is a product of the USIA itself, and the second comes out of the United States Department of Defense. His argument is seriously undermined by his use of the same material generated by the very American agencies he criticizes for failing to build a cultural bridge with the Vietnamese. He contends that Americans failed to cope with the "Confucian passivity" that they found "persisting in the South Vietnamese countryside," but the only evidence he produces to indicate that this situation proved to be the case are assessments offered about Vietnamese character that the USIA produced.[9]

Chandler's material is much more valuable for understanding how Americans perceived Vietnamese than it is for the way he uses it to generalize about Vietnamese culture, but he elects not to explore that avenue. Instead, he references a 1965 USIA report and states as a matter of fact, "The South Vietnamese peasant views the world outside his village as being beyond his control; he finds little reason to concern himself with or take part in national affairs." In Chandler's view, "It is not surprising that so many Vietnamese over the years failed to side firmly with either Hanoi or Saigon."[10] Nevertheless, he fails then to explain, with so many Vietnamese supposedly on the sidelines, where all of the insurgents and their supporters came from.

Questions must also be raised about the measuring devices Chandler uses to gauge the effectiveness of America's effort. He asks, "On balance, how did the Vietnamese people judge Saigon's argument?" For his answer, Chandler turns to "one American propagandist" who tells him, "Fundamentally pacification was a success. . . . With the new mobility and communications we brought here, the Vietnamese people now are much more sophisticated and much less susceptible to enemy propaganda techniques."[11] Without returning to the difficulty contained in the act of evaluating PSYOP, merely restating an American participant's impression of how the program affected Vietnamese alleviates few of the problems and generates little insight about the operation's real value. It only underscores the impressions Americans had already formed about the relative status of Vietnamese to themselves and about the validity of technique as a substitute for ideas and action, the same type of thing that occurred so often during the war.

Mirroring the assessment of the material he draws from that which the USIA had generated during the war, Chandler constructs the Vietnamese as a group of people who largely remained apathetic about the war raging around them. He theorizes that they stood "preconditioned" to accept whatever the fates brought them, and he alleges that "most assumed a 'wait and see' outlook and remained uncommitted to either side throughout the war."[12]

Chandler's work, and those individuals who endorse it, lend credence to the idea long used by American officials to rationalize the difficulties their PSYOP program encountered in Vietnam. That is, Americans, despite their best intentions and hard work, could not help a group of people who proved unwilling to help themselves. According to this view, Americans did not succeed in winning Vietnamese hearts and minds because they could not overcome a culture of apathy and disengagement. Rather than serving as analysis, it only perpetuates what American officials believed throughout the war about passive Vietnamese being manipulated from the outside. The problem here, of course, is that it neglects to consider the expressions of discontent against Saigon officials that originated from the people living in the southern half of Vietnam.

Chandler is most successful with his study in indicating the level of consistency, an attribute he considers one of the most important for effective PSYOP, that the JUSPAO achieved with its communications. By qualitatively comparing over one thousand of America's leaflets, Chandler finds that the JUSPAO achieved over ninety-five percent compatibility with the messages it delivered in both the DRV and the RVN. He concludes, "Indeed, given the complexity of the insurgent environment and the long chains of coordination between the military and pacification aspects of the propaganda campaign, there was a surprising extent of consistency." The "overwhelming compliance" of the program, he argues, "appears to be a plus for possible effectiveness."[13]

Chandler's inquiry into the consistency of America's communications is important. Consistency goes a long way toward establishing credibility with an audience, a status that could have undoubtedly helped Americans make their case in Vietnam. Important to credibility beyond mere verbal uniformity, though, is the ability to harmonize actions and words. While comparing American messages against one another, it eludes Chandler that one must gauge whether the words utilized by the United States accurately reflected the activities it pursued as well. Americans could uniformly broadcast to forces opposed to the RVN, for example, that any person who defected would receive fair and humane treatment. Still, the message would fail to have its intended effect of seducing more defectors if concrete actions did not ensure those who set down their arms actually received the reception promised by leaflets. This inconsistency between rhetoric and action proved a recurring problem for the defector program that surely worked against leaflet-made promises no matter how regularly they appeared, but it goes unacknowledged by Chandler's study, which focuses on what the JUSPAO promised Vietnamese at the expense of what the agency actually managed to deliver them.[14]

Throughout his work, Chandler not only makes the mistake of rendering judgment about the program's effectiveness wholly from an American participant's perspective based on official sources, but he also fails to consider adequately the totality of PSYOP. He repeats the process followed by United States officials at the time by divorcing the information program from American-RVN conduct during the war and by assuming Vietnamese represented little more than blank slates for either Americans or communists to write on. Chandler encourages one to look at the program as if its operations consisted of an array of radio broadcasts and leaflets that reached passive subjects in a vacuum tube as opposed to people actively caught up in the events of their time. He does not scrutinize the message Americans conveyed in Vietnam against any conditions that existed there. Throughout his work, Chandler is far less concerned with what Americans said than he is with how they said it. Moreover, he neglects entirely to consider that Vietnamese had anything to say at all.

LASSITUDE WILL BE LASSITUDE

To the extent students of America's participation in the Vietnam War have given the subject any thought, they overwhelmingly endorse the interpretation of PSYOP Chandler offers in his work. In his survey of Vietnam War literature, eminent historian Gary R. Hess credits Chandler with suggesting the "enormity, if not the impossibility, of the American challenge in the countryside." He acknowledges the validity of Chandler's conclusion that America's results proved "disappointing" and that the United States "propaganda campaign failed precisely because it was an American, not a Vietnamese, program."[15] Hess's comments, noteworthy for their endorsement of Chandler's thesis, also stand out for their uniqueness, as the entire topic of PSYOP in Vietnam is a grotesque lacuna in the literature.

In fact, one must mine deep merely to find any mention of America's attempts to communicate persuasively with Vietnamese. American Studies scholars Jon Roper and Phil Melling, in *America, France and Vietnam: Cultural History and Ideas of Conflict*, a collection of papers presented in 1988 at the American Studies Centre at the University of Wales, Swansea, address the issue. While they do not cite Chandler's work, Roper and Melling contend, "Given the cultural and ideological assumptions which formed the basis of their perspective on the wider world, there was little prospect of dialogue either between Americans and the nation they tried to build in South Vietnam, or Americans and the people they proceeded to fight in North Vietnam."[16]

Support for this assessment is also widely accepted, if rarely cited, by individuals who served in the USIA during the war. Fitzhugh Green labels USIA efforts in Vietnam an "ultimate disappointment." He recognizes his former colleagues as "dedicated, industrious, and often ingenious," but Green laments that their initiatives "were doomed to a half-baked, almost indistinguishable result." He echoes Chandler's contention by proposing that the JUSPAO never stood a chance for success. He writes, "How could they [Americans] communicate through the opaque wall of language and culture? Almost no one in JUSPAO ever breached it."[17] Even former JUSPAO director Barry Zorthian questions whether or not the United States encountered an entirely unworkable situation in Vietnam. Zorthian observes that America's "basic experience in conflict usually involved civilian populations of ethnic and cultural background that had some mutuality and relevance to our [American] own experience." Earlier successes the United States had mustered in conveying its message abroad occurred with people who possessed a "similar Judeo-Christian tradition and, thus, with roughly the same ethics and values."[18]

According to Zorthian, Americans went to Vietnam woefully ill equipped to communicate with the people they found there. The Vietnamese, the former JUSPAO director and his old colleagues believe, shared fundamentally little in common with Americans. Zorthian contends that the United States possessed "no experience relevant to Vietnam from the viewpoint of PSYOP." The nation then compounded its situation by sending Americans into the Vietnamese countryside who learned about their new surroundings through training that "wasn't very deep or very thorough or very complete." Once in the field-Americans had no choice, according to Zorthian, but to rely on "conventional wisdom anyone would get when they got in the field about the Confucian mind and oriental thinking, Vietnamese."[19] For Zorthian and other JUSPAO personnel, this "conventional wisdom" drew from a deep well of stereotypes Americans had long held that posited "oriental thinkers" as technologically backward and stubbornly resistant to innovation and change.[20]

That Americans considered their traditional ideas about "oriental thinkers" sufficient for the task ahead of them is obvious when one considers that they had already been in Vietnam for over a decade before Zorthian arrived in 1964, a more than reasonable amount of time for them to accumulate some sort of understanding about the people they ostensibly had come to help. They neglected to do so, however, and after the war ended they continued to hold fast in their beliefs that when presumed American initiative mixed with apparent Vietnamese apathy the results stood as foreordained.

Zorthian proposes that differences in "pace and concepts, in ethics and customs, in desire and capability, led Americans to conclude repeatedly—and perhaps rightfully—that the Vietnamese were not doing the job." He writes, "As a result, proverbial American impatience led to the inevitable effort to do the job ourselves. The effort was doomed to failure before it started."[21] Harry Latimer, who served as the senior military officer on loan to the USIA as Chief of Plans, Policy and Research in the JUSPAO, concurs. He concludes, "A very great error was made by the U.S. in trying to do the job for the Vietnamese."[22]

Contained in these statements is the notion that the Vietnamese proved either incapable of or unwilling to demonstrate any independent initiative. They are the ones who stood as culpable for whatever difficulties Americans experienced. Latimer makes the idea explicit. He writes, "The lack of competence of VIS was for most the primary frustration, for if it did not appear the Vietnamese Government could handle matters after U.S. withdrawal, why all the effort?"[23] The component lacking from these appraisals is the willingness to question whether the PSYOP program mistakenly conflated Vietnamese culture as a whole with the particular RVN regime Americans supported because it appeared to fit the description they had already devised. American officials came to conceive in Vietnam that "lassitude will be lassitude" among Vietnamese, and that the war's psychological terrain would have to be conquered the American way, a process that paved the way for others credibly to indict the United States on charges of neocolonialism.[24]

Americans saw no evidence of initiative within the government that the United States sought to establish with legitimacy in Saigon, assumed none existed throughout the country, and proceeded to take it as their job to make the individuals they backed at least "look inspired" for the benefit of gaining international support. A remarkable testament to exactly how disconnected Americans remained from the history of the nation where they now intervened, they believed it more likely that their effort could only foster the appearance of, rather than the actuality of, Vietnamese self-government. Americans had made their own the "charade" concocted by the French who, prior to their departure in 1954, claimed not to fight a colonial war against Vietnamese nationalists but to support one contestant engaged with another in a civil war.[25]

This mentality, the denial that Vietnamese nationalism constituted an indigenous force that took inspiration from Vietnam's history of independent government, enabled the United States to engage in the task it described as "nation building" in the southern part of the Vietnamese nation. Americans maintained Vietnamese lacked a national identity; instead, they insisted that:

> One aspect which pervades much of Vietnamese psychology is *regional* origin—whether a Vietnamese considers himself from the *North* (primarily Hanoi), the *Center* (Hue to Nha Trang) or a *Southerner* (Saigon and the Delta), or else a member of minority groups.[26]

Americans noted this "psychological orientation" of Vietnamese that they identified as possessing its "roots in the long centuries before 1954 of the three civilizations of the Tonkin Delta, Central Annam and Cochin China." According to American officials, the legacy of these "historical civilizations" meant, "for many Vietnamese, a close, emotional affinity with their region often underlies their cultural and political appraisal of other Vietnamese."[27]

This assessment, while convenient for Americans who detected the need to build a nation in southern Vietnam, is at odds with Vietnamese history. As political scientist Robert Scigliano managed to observe after spending two years in Vietnam as a member of the Michigan State University Vietnam Advisory Group (1957–1959), the regional psychology of Vietnamese that Americans identified resulted not from Vietnam's past but from French imperial design. Scigliano proposed that "French policy in Vietnam denied the very existence of the Vietnamese nation." For the French, because it made good administrative sense to them for their colonial purposes, there only existed "the Tonkinese, the Annamese, and the Cochinchinese," the same groups that Americans would later recognize as well. According to Scigliano, though, it constituted a "serious mistake not to recognize the fundamental unity underlying the apparent diversity within Vietnam."[28]

Scigliano had come to understand better than most Americans that Vietnamese history extended back in time past the time of French colonization (1874–1954) and could be distinguished throughout by a profound national consciousness. The ancestors of modern Vietnamese had created an organized society based in the Red River Delta called Van Lang by the middle of the second millennium B.C.E. Although it was conquered by the Han Chinese by the end of the first century C.E., the memory of independence preserved a sense of cultural identity that enabled Ngo Quyen to lead a successful insurgency against Chinese rule in 938 C.E.

Over the course of the next seven centuries, a succession of Vietnamese dynasties managed to create a centralized bureaucratic structure that constructed a postal system, built a network of roads, and employed civil service examinations and military conscription. A concurrent expansion to the south by the Vietnamese state brought the Vietnamese absorption of the once mighty Empire of Champa's territory as well as the southern plains of modern Vietnam from the Khmer by the end of the eighteenth century. During this period of southern expansion and bureaucratic centralization, the Le Dynasty (1428–1788) provided Vietnam with nominal rule, while rival families, the Nguyen in the south and the Trinh in the north, competed with one another for real control using military means.

By the beginning of the nineteenth century, following their defeat that gave rise to the short-lived Tay-son Dynasty (1788–1802), the Nguyen family reestablished its might and restored unity. Nguyen Anh assumed the title of Gialong (1802–1820) and established a capital at Hue for the kingdom he proclaimed would be known as Viet Nam.[29] After nearly two hundred years of intermittent civil war, Vietnam stood as reunified, and its people's sense of national identity clearly remained intact. As one French officer noted later in the nineteenth century about his nation's latest colonial addition, "We found ourselves, from an ethnic as well as a political and social viewpoint, in the presence of the most unified people one could imagine."[30]

This lesson learned by at least some of the French who went to Vietnam went largely unrecognized by Americans during and after the war. In 1964, USIA director Carl Rowan and other American officials drew up a blueprint to prevent the status of Vietnamese nationalism as they defined it, allegiance to the RVN, that they saw as already "grim" from becoming "unrecoupable." They devised an American-style media blitz, replete with films, televisions, and radios to be distributed throughout the countryside. They planned "fireside chats" to be held by RVN leaders, they considered allotting a "cash gift or piece of land" for stories of individual heroism, and they even called for the need to create a Vietnamese version of "God Bless America."[31]

Rowan's plans assumed Vietnamese needed to be injected with a nationalist spirit from the outside, that they lacked the capacity to create a nation without American assistance. It never occurred to Rowan that the nation he and other Americans wished to create for Vietnamese might run counter to the historic notion of nationhood already in place throughout Vietnam, one that lacked an artificial dividing line along the seventeenth parallel that had been imposed from the outside.

The program that the USIA conceived emphasized the need to create patriotic imagery and operated on the assumption that the Vietnamese lacked a historical national identity. It received validation from western Asian Studies scholars that cast post-colonial societies like Vietnam as akin to infants that needed to receive a tutorial in western ways. William L. Holland, for example, who served as general secretary and research director of the Institute of Pacific Relations (1946–1960) suggested that the "ordinary villager" in Southeast Asia must feel "that the nationalist governments are almost as remote and alien" as "were the former colonial regimes." He, therefore, identified as "an urgent task" the need to "deliberately and quickly" create "new symbols, slogans, precepts, patriotic ceremonies, by which the masses can more readily identify themselves with their new government."[32]

Looking back at the USIA program he initiated in Vietnam, Rowan raises no questions about the content of the campaign Americans constructed or even about the means they chose to carry it out. Instead, he wonders only about the wisdom of Americans who considered it possible to motivate the Vietnamese in the first place. Rowan concludes that he and his fellow propagandists had "erred" in believing they could "mount some propaganda campaign" that would impel Vietnamese to denounce "political slavery, slaughter," and "tyranny."[33] Rowan's postwar admission primarily concedes that he overestimated the ability of PSYOP alone to counteract the lingering effects Vietnam felt from the near century of foreign oppression it had endured. It also demonstrates, however, the depth of his belief Vietnamese stood as unable independently to shed the evils that confronted their society, and they lacked the necessary maturity as a people to grasp assertively the assistance America offered them.

Rowan utilizes language that conveys the conviction he shared with other Americans who held, that even in the face of what they saw as wanton oppression, Vietnamese people proved unable to take a stand because it went against their presumably listless nature. Rowan calls his plea for increased American PSYOP in Vietnam a "mistake" only multiplied by French, Vietnamese, and United States officials. Given the "arrogance" and the "shameless crookedness" evident in Vietnamese history, the "colonial arrogance" of the French, the "venality and incompetence of those who at the time held power in Vietnam," and the "excuses" emanating from the Johnson administration, Rowan asks, "How could any propaganda campaign succeed?" In one brief passage, Rowan blames everybody involved in the communications endeavor for the failure of Americans to connect with Vietnamese except for United States information officials.

While it is clear the program that the USIA directed through the JUSPAO encountered difficulties from all sides, it is also apparent PSYOP officials rarely considered incorporating into policy the input their program solicited in any way that remotely resembled its capacity to disseminate American-generated material. PSYOP officials detected problematic aspects of both RVN rule and the American presence in Vietnam but often ascribed them to the product of Vietnamese nature. Other American decision-makers demonstrated little responsiveness to the observations that JUSPAO field officers did make. In most cases, American officials professed their realization of the need to win Vietnamese support, but they did little to connect action with rhetoric.

Americans repeatedly exhibited the most interest in pacifying the population at large and creating the impression of a stable RVN-ruled society by channeling the bulk of the agency's relatively meager resources into securing good relations with the media gathered in Vietnam. Officials like Rowan, both during and after the war, confused the Vietnamese population as a whole with the inept government officials America strove to keep in place. They have done little, however, to question the role that the PSYOP program played in facilitating the support America provided to the RVN in the first place. Even with the hindsight he possessed at the time he wrote his memoirs, Rowan never reconsiders the legitimacy of American assumptions regarding Vietnamese apathy.

An additional problem with Rowan's assessment, and with the whole memoir-advanced idea that the American effort stood as doomed from the start due to Vietnamese inertia, is that by the most important measure, the war's ultimate outcome, the techniques used by Vietnamese communists did succeed. Opposition to the RVN garnered enough support ultimately to topple the government in Saigon. Communism constituted no more of an ideology indigenous to Vietnam than did the ideas Americans sought to disseminate there, but those who propagated it succeeded in communicating it to people as the most viable version of Vietnamese nationalism available. There is little reason to suspect a communist form of nationalism would have had a greater inherent appeal to Vietnamese than that held by American propagated ideas of capitalism and democracy. Both sets of ideas originated elsewhere, but only the American-supported notion lacked what the Vietnamese perceived as a legitimate historical component born out of their nation's historical experience.

A CULTURAL BOX

Besides the memoir accounts, other studies that employ a cultural prism to place in context the whole of America's effort in Vietnam also suggest Americans had little chance for success in crossing the great cultural divide they encountered in Vietnam. Works by the journalist Frances Fitzgerald and historian Loren Baritz are the two most noteworthy examples. Baritz cites a 1974 survey conducted among United States Army generals to illustrate American ambivalence regarding Vietnamese culture. He notes that one of the questionnaire's respondents lamented, "we never took into account the cultural differences." Another commented, "as in all our foreign wars, we never really established rapport."[34] While Baritz calls attention to America's inability to overcome cultural obstacles, Fitzgerald goes even further with her criticism, alleging Americans simply did not care about Vietnamese at all.

In her study, Fitzgerald accuses United States officials of "invincible arrogance," charging that they remained relatively unconcerned about the Vietnamese. She indicts American policy for overlooking morality, an aspect she refers to as "the heart of the matter," and for concentrating coldly on only its own strategic interests. Fitzgerald then goes on to chronicle the cultural differences she sees that "made the whole enterprise, from the most rational and tough-minded point of view, risky in the extreme."[35] The implication here is had United States officials approached Vietnam in a more pragmatic fashion, they would have avoided the whole endeavor in the first place by concluding Vietnamese simply could not be made into good Americans. It is a problematic assertion because it echoes American thinking during the war by positioning the United States further ahead on some imagined evolutionary scale where Vietnam lagged hopelessly behind. Like American officials who retrospectively question the ability of anybody to help the Vietnamese, Fitzgerald blames United States leaders for not recognizing Vietnam's static and child-like state.

Fitzgerald asserts, "In going into Vietnam the United States was not only transposing itself into a different epoch of history; it was entering a world qualitatively different from its own." Interestingly, she uses the term "world" in her analysis rather than a word like either "nation" or "society," suggesting Americans and Vietnamese did not even inhabit the same planet. To learn about one another, she suggests, these people, who she implies might have well been otherworldly aliens to one another, "would have to re-create the whole world of the other, the whole intellectual landscape." From that point, she contends, Americans would have to wrestle with questions of morality, to "consider which of their values belong only to themselves or only to a certain stage of development."[36]

FitzGerald's reference to evolutionary stages reveals her willingness to tap into the same beliefs held by American officials at the time, that societies of people pass through some sort of linear progression and that the United States clearly exceeded Vietnam's station on the line. Walt Whitman Rostow is a former OSS operative during World War II, who went on to become a professor of economic history at the Massachusetts Institute of Technology, he authored "Non-Communist Manifesto" in 1960. It chronicles the possibility of "progress" for the "whole pre-Newtonian world" provided conditions are established that allow "traditional societies" to "take-off" and "drive to maturity."[37]

Rostow went on to become an important member of the presidential administrations of both John Fitzgerald Kennedy and Lyndon Baines Johnson.

Kennedy named Rostow the Deputy Special Assistant for National Security Affairs, which he acted as briefly in 1961 before becoming counselor of the Department of State and Chairman of the Policy Planning Council, Department of State. In 1964, Rostow wrote President Johnson's first State of the Union address, and he accepted a presidential appointment to act as the United States Member of the Inter-American Committee on the Alliance for Progress (CIAP). Johnson then named Rostow Special Assistant for National Security Affairs (now referred to as National Security Advisor) in 1966. Rostow remained at this post until February 1969, and he played a leading role in developing the Johnson administration's policies in Vietnam. Rostow's ideas that posited the world's societies in relation to one another according to their "stages of evolution" differ little from Fitzgerald's later analysis.[38]

FitzGerald portrays Vietnamese as polar opposites of Americans with grand and sweeping generalizations. Citing their differing sources of inheritance, China for Vietnam and Europe for America, she writes, "In their own independent development the two nations have in many respects moved even further apart from each other." She sees American culture as a product of "conquest and expansion" and the Vietnamese way of living as chiefly characterized by a "sense of limitation and enclosure." This difference, she suggests, left two societies that "stand in the relationship of a reversed mirror image, for the very notion of competition, invention and change is an extremely new one for most Vietnamese." While Americans "canted towards the future, the traditional Vietnamese were directed towards the past."[39]

Like the reminiscences of former USIA personnel, the central message of Fitzgerald's study is that the American effort in Vietnam stood no chance from the beginning because fundamental differences separated the two groups of people, and United States officials failed to recognize this gap as significant to their designs. Whether they failed to consider it, or if they simply dismissed it, the premise is that American planners did not incorporate culture into their considerations for Vietnam, and they wrongfully assumed Vietnamese capable of instituting their own democratic government. In Fitzgerald's view, Americans had mistakenly placed their concerns for "the containment of Communism and the security of the Free World to a dimension where governments rose and fell as a result of arguments between two colonels' wives."[40]

Culturally rooted explanations such as Fitzgerald's offer valuable insights. Perhaps most importantly, they reinforce the importance of cultural considerations to international relations. Akira Iriye observes, while nations often act to enrich themselves, or to protect themselves, "they also develop visions, dreams, and prejudices about themselves and the world that influence their relationships."[41] No less than economic, political, or territorial concerns, culture must be accounted for in analyses of global affairs. As Iriye suggests, the dreams and visions privileged by leaders can play an important role in determining the policies that a nation prioritizes, but they can also prove seductive to the scholars that study them. Baritz and Fitzgerald demonstrate this point as, undoubtedly inspired by the insights that culture offers, they repeatedly offer their readers careless generalizations. For example, while FitzGerald's commentary about the impact made by the arguing spouses of officials might point to the particular deficiencies of those individuals who administered the RVN, it is careless of her to suggest Vietnam stood capable of no better than a government that ruled according to whim. Vietnamese opposition to the RVN stemmed at least in part to just such arbitrary exercises of power.

Fitzgerald and Baritz treat culture as a static component of society rather than as an organic attribute. They allow for neither change nor variation in the values they define; instead, they describe monolithic Vietnamese and American national mindsets that move each group of people inexorably down a predetermined path that leaves no room for deviation. Such analysis concocts imaginary national psyches that accommodate neither internal variation nor the possibility of change over time. It should be obvious that no group of people acts in harmony at all times, or that not always does a single person's behavior subscribe consistently to a particular set of ideals. Otherwise, the gap between ideal and practice is obscured. To categorize entire groups of people as either forward looking or backward glancing is simplistic at best, racist at worst. It exaggerates difference and diminishes similarities between people, and it reduces the behavior of entire societies to the product of a cultural stereotype.

In his review of Fitzgerald's book, David G. Marr, a Vietnamese studies scholar and a former intelligence officer with the United States Marine Corps in Vietnam, refers to the author's characterization of Vietnamese national character as "little short of disastrous." He astutely points out the generalizations contained in *Fire in the Lake* would be "risky business," even if the author evidenced greater familiarity than she does with either Vietnamese language or literature.[42] More recently, Robert Templar, Asia program director, International Crisis Group, has criticized Fitzgerald's portrayal of Vietnamese culture as little more than a caricature. He writes:

> Fitzgerald's Vietnamese are devoid of all diversity or individuality; they remain trapped in fixed intellectual and physical landscapes, completely beholden to the ageless and unbending forces of Confucianism, colonialism and village life.[43]

Templar's work directly challenges the imagery conjured by Fitzgerald's interpretation. He makes it clear that Vietnamese culture is more dynamic than portrayed in *Fire in the Lake.* "Far from being stuck in their villages, tending their walls of bamboo," he shows, "Vietnamese have for centuries been traveling, trading, migrating, conquering, fleeing, expanding and exploring."[44] This view is supported by the exploration of Vietnamese history and society offered by Mark W. McLeod and Nguyen Thi Dieu in *Culture and Customs of Vietnam.* While this work makes clear that on the eve of French colonization in the nineteenth century, Vietnam remained overwhelming agricultural, most farming families engaged in some type of non-agricultural trade at least some of the time. In addition, although careful not to over exaggerate its success, the authors illustrate that the Nguyen dynasty that came to power in 1802 had initiated a process of centralization that stimulated the development of administrative centers around the country that emanated from the capital city of Hue. They also make clear that, by the early nineteenth century, Saigon had emerged as a busy port for international trade.[45]

The Vietnamese culture that Fitzgerald depicts, and that she accuses American officials of ignoring, in fact appears quite similar to how American PSYOP projected Vietnamese society during the war. The JUSPAO routinely churned out characterizations of all facets of Vietnamese life for the consumption of United States officials that stand as at least as sophisticated as the assessments rendered by critics such as FitzGerald, who accuse the United States of ignoring these very same characteristics. Often they are strikingly similar, constructing Vietnamese values in exact opposition to those ideals perceived by Americans as unique to themselves and placing the United States measurably in front of Vietnam on some conjured Darwinist evolutionary line. Moreover, they are notions that strongly resemble the ones propagated about the Vietnamese by the French.

Paul Mus, a French scholar who served as the political adviser to the commander of France's reoccupation forces in Indochina, General Philippe Leclerc, became a leading proponent of these views in America during the 1950s as a professor at Yale University. *The Vietnamese and Their Revolution*, which Mus co-authored with John T. McAlister Jr., is a based on Mus's *Viet Nam: sociologie d'une guerre*, published in 1952. In it, the authors describe the Vietnamese as occupants of a land that time forgot. They observe, "After going mile after mile in the Vietnamese countryside one, has the impression of having stood still."[46]

A 1966 JUSPAO memo asserts, "The American—whose forebearers tamed a continent, whose scientists daily perform miracles in outer space—seeks to alter the word to suit his needs." It then proceeds to make the same point made by critics such as Fitzgerald, who point to cultural unawareness as the problem with America's policy in Vietnam. The memo notes:

> What Vietnamese see as suppleness we see as duplicity, what they see as tranquillity of spirit we see as lack of concern. And what we see as resolution the Vietnamese see as intractability, what we see as boldly meeting challenge they see as disrupting of the harmony of the universe.[47]

Far from ignoring Vietnamese culture, Americans categorized it to their satisfaction and decided it insufficient for the task they identified on the horizon. JUSPAO officials started with the assumption that their mission constituted providing Vietnamese with the inspiration for self-determination they inherently lacked. They then went on to define Vietnamese values through direct comparison with how they perceived American ideals, literally framing the relationship in the context of "us and them." This strategy resulted in Americans viewing Vietnamese in a way that starkly contrasted to the way they interpreted their own world-view.

Americans typically saw Vietnamese flailing about according to whichever way the wind blew, while they considered themselves as rushing headfirst into whatever the maelstrom to affect change for the better. Americans quickly placed Vietnamese into a tightly sealed cultural box where they would keep them through the duration of the war. By examining the history of United States PSYOP prior to Vietnam, it becomes clear that Americans had begun constructing this container long before they ever arrived in Southeast Asia.

NOTES

1. Charles R. Anderson, *Vietnam: The Other War* (Novato: Presidio Press, 1982); Chandler, *War of Ideas*; Fitzhugh Green, *American Propaganda Abroad* (New York: Hippocrene Books, 1988); Latimer, *US Psychological Operations*, 1973); Mecklin, *Mission in Torment*; Rowan, *Breaking Barriers*; Sorensen, *The Word War*; Zorthian, "The Use of Psychological Operations."
2. Latimer, *US Psychological Operations,* 49.
3. See Kahin, *Intervention*, 93–121; Kolko, *Anatomy of A War*, 92–96.
4. Chandler, *War of Ideas*, 11.
5. Chandler, *War of Ideas*, 10.
6. Chandler, *War of Ideas*, 11.
7. Chandler, *War of Ideas*, 16.
8. Chandler, *War of Ideas*, 20.
9. Chandler, *War of Ideas*, 16.
10. Chandler, *War of Ideas*, 17.
11. Chandler, *War of Ideas*, 205.
12. Chandler, *War of Ideas,* 10.
13. Chandler, *War of Ideas*, 228–30.

14. Report, Improving the Effectiveness of Chieu Hoi, September 1967, "Vietnam: Simulatics Corporation: Improving Chieu Hoi," NSF—Country—Vietnam, box 239, LBJL.

15. Gary R. Hess, "The Unending Debate: Historians and the Vietnam War," in *America in the World: The Historiography of American Foreign Relations Since 1941*, ed. Michael J. Hogan (Cambridge: Cambridge University Press, 1995), 388.

16. Phil Melling and Jon Roper, eds., *America, France and Vietnam: Cultural History and Ideas of Conflict* (Brookfield, Vermont: Gower Publishing Company, 1991), 26.

17. Green, *American Propaganda*, 151.

18. Transcript, Barry Zorthian interviewed by Ted Gittinger, May 26, 1982, 84–64: 12–13, Oral History Collection, LBJL.

19. Transcript, Barry Zorthian interviewed by Ted Gittinger, May 26, 1982.

20. Jacobson, *Barbarian Virtues*, 29–38.

21. Report, Conduct of the War: Functional Analyses, Volume VI Book 2 [3 of 3], May 2, 1980, "Conduct of the War: Functional Analyses," box 24, BDM Corporation Study: A Study of Strategic Lessons, LBJL.

22. Latimer, *US Psychological Operations,* 50.

23. Latimer, *US Psychological Operations*, 43.

24. "Summary Record of Meetings, Honolulu," June 2, 1964, *FRUS, 1964–1968*, I: 428–430.

25. Kahin, *Intervention*, 26.

26. Booklet, *JUSPAO Vietnam*, 6 (emphasis in original).

27. Booklet, *JUSPAO Vietnam*, 6.

28. Robert Scigliano, *South Vietnam: Nation Under Stress* (Boston: Houghton Mifflin Company, 1963), 6.

29. Neil L. Jamieson, *Understanding Vietnam* (Berkeley and Los Angeles: University of California Press, 1993); Mark W. McLeod and Nguyen Thi Dieu, in *Culture and Customs of Vietnam*, Culture and Customs of Asia series, ed. Hanchao Lu (Westport, Connecticut, and London: Greenwood Press, 2001); Keith Weller Taylor, *The Birth of Vietnam* (Berkeley and Los Angeles: The University of California Press, 1991).

30. Quoted in Scigliano, *South Vietnam*, 7.

31. Scigliano, *South Vietnam*, 429.

32. William L. Holland, "Asian Nationalism and the West," in *West and Non-West: New Perspectives*, ed. Vera Micheles Dean and Harry D. Harootunian (New York: Holt, Rinehart and Winston, Inc., 1963), 168.

33. Rowan, *Breaking Barriers*, 263.

34. Baritz, *Backfire*, 25.

35. Fitzgerald, *Fire on the Lake*, 7.

36. Fitzgerald, *Fire on the Lake*, 7.

37. W. W. Rostow, *The Stages of Economic Growth; A Non-Communist Manifesto* (Cambridge: Cambridge University Press, 1960).

38. Rostow, *The Stages of Economic Growth.*

39. Fitzgerald, *Fire on the Lake*, 5.

40. Fitzgerald, *Fire on the Lake*, 5.

41. Akira Iriye, "Culture and International History," in *Explaining the History of American Foreign Relations*, ed. Michael J. Hogan and Thomas G. Patterson (Cambridge: Cambridge University Press, 1991), 217.
42. David G. Marr, "Fire in the Lake: The Vietnamese and the Americans in Vietnam," *Journal of Asian Studies,* 32, no. 3 (May 1973): 564–65.
43. Robert Templer, *Shadows and Wind: A View of Modern Vietnam* (New York: Penguin Books, 1999), 18.
44. Templar, *Shadows and Wind*, 18.
45. McLeod and Nguyen, *Culture and Customs of Vietnam*, 22–26.
46. John T. McAlister Jr. and Paul Mus, *The Vietnamese and Their Revolution* (New York, Evanston, and London: Harper & Row Publishers, 1970), 51.
47. Memo, JUSPAO Field Memorandum Number 27—The Older Vietnamese as Communicant, October 4, 1966, "CI File," Lee Papers, MHI.

Chapter Four

Lessons Learned

OVERVIEW

This chapter locates the JUSPAO that Americans created in Vietnam within the context of America's previous experience in conducting PSYOP during the twentieth century. It is divided into three sections, with the first entitled "The Importance of Hearts and Minds." This initial part of the chapter underscores the intentional separation made by Americans during the First World War to draw a line between information activities and psychological warfare.

The chapter's second installment, "Nothing but the Truth," illustrates that during the Second World War Americans made the boundary even more firm than it had been between their information and propaganda agencies, in part because of the domestic criticism that developed during the First World War about America's operation of an information program. The notion that American information differed from propaganda by its "objectivity" and its "truthfulness" became increasingly more accepted.

"Bats and Foxes, Oh My!" the concluding section of this chapter, illustrates that, during the Second World War, Americans identified pronounced psychological differences between their European and Asian enemies. They deemed material appropriate for the Japanese that they dismissed outright for use against Nazi Germany and established an important precedent for the American conduct of PSYOP in Vietnam.

THE IMPORTANCE OF HEARTS AND MINDS

"What we've got here is . . . failure to communicate." The phrase famously uttered by the prison captain to the inmates whose work he supervised in

Hollywood's 1967 film *Cool Hand Luke* is easily lent to describe the situation that developed in Southeast Asia even as the movie played in American theaters. At the very least it is an assessment consistently offered by both individuals who participated in America's involvement in Vietnam and those who have subsequently considered the war. The standard refrain plays: Had Americans only been able to relay their intentions successfully and to breach the great cultural divide they found occupying the space between themselves and Vietnamese, the United States would surely have won over to its side the hearts and minds it so desperately coveted. This line of reasoning concludes by suggesting that since Americans could not connect with their intended audience, the Vietnamese must have been unreachable.

Real differences exist between Vietnamese and American culture, but this study posits that these dissimilarities alone did not prevent the United States from reaching people. In fact, it holds that Americans made their message quite clear regarding Vietnam in many ways; moreover, their delivery could not have been more convincing, cultural differences notwithstanding. The message that Americans sent undeniably resonated with many Vietnamese, simply not in the way United States officials intended. Don Luce noted in 1972, after spending thirteen years in Vietnam as an agriculturist and as the director of International Voluntary Services, "The United States has made more 'Viet Cong' than it has killed."[1] His observation is echoed by comments made by Vietnamese such as Phan Van Hoang, a student in Hue in 1968 who attributes his enlistment into the People's Liberation Army (PLA) to his encounter with the United States Central Intelligence Agency (CIA). "I had been a student activist before that time, not a guerrilla," he remembers. "But then my experience with the Americans changed all that."[2]

Ostensibly, Americans stood well prepared to communicate their policy persuasively in Vietnam and to win allies rather than foster opposition. By the time the first USIS officers arrived in Vietnam in the early 1950s, the United States had been engaged in courting world opinion by using government information agencies to support its policies for almost four decades that included two successfully waged world wars. In addition, the nation's leaders possessed a resounding confidence in the idea that persuasive communications represented a singular American genius. In 1950, United States President Harry S. Truman proclaimed, "We have tremendous advantages in the struggle for men's minds and loyalties. We have truth and freedom on our side."[3] Truman's voice found an echo in Barry Zorthian's two decades later, when the JUSPAO director noted while reflecting on his experience that the entire endeavor of PSYOP resided in an area where "Americans should logically excel." After all, he asked, "Who else made the 'opinions of mankind' such a tent of faith?"[4]

In one way, the JUSPAO organization constituted a new approach for Americans in seeking to win the endorsement of others for their policies. Unlike previous forays in the realm of international opinion, the United States combined two different elements of PSYOP, information and psychological warfare or propaganda, under one roof. Historically, Americans had kept these activities distinct from one another and under the direction of different organizations to avoid accusations of targeting civilians with propaganda. During the First World War, psychological warfare, then called propaganda by American officials, fell under the purview of the United States Army's Propaganda Subsection, G-2-D.[5]

Created on June 21, 1918, under the United States War Department and commanded by Captain Heber Blankenhorn, who worked as a journalist in New York City prior to the war, G-2-D carried out tactical campaigns of a strictly military nature aimed at securing immediate results. Primarily, Blankenhorn's unit concentrated on dispersing leaflets aimed at seducing as many enemy defections as possible. At war's end, the outfit had saturated Europe with over five million leaflets that it distributed from the rear areas surrounding Metz to the Argonne Front. By the time G-2-D became operational in August 1918, America's first official information agency had been up and running for over a year. United States President Woodrow Wilson created the Committee on Public Information (CPI) with his friend and staunch political supporter, Colorado newspaper editor George Creel, as its director.[6] Creel had authored the partisan *Wilson and the Issues* to help further the political fortunes of his friend in 1916, and his wife had long maintained close ties with Wilson's daughter Margaret.

For a variety of reasons, Wilson and Creel chose to use the term "information" to label their new agency's activities as "informational" as opposed to "propaganda." According to Creel, they made their selection because America's enemies, Germany and Austria-Hungary, had created an association between "propaganda" and deceit.[7] Other factors, though, undoubtedly played a role as well. America's western allies both already had created their own information departments. The French formed the Press and Information Bureau in August 1914, and the British had created the Ministry of Information in early 1917.[8] The emphasis these agencies placed on countering what they called enemy lies could not have escaped Wilson and Creel. They not only wished to use the CPI to sell America's war aims to the rest of the world, they also needed to use it to convince their own public of the righteousness of its nation's participation. They needed to diffuse the notion, especially because of their close relationship, that the CPI advanced the interests of the United

States and not just the presidential agenda of Woodrow Wilson. The rhetoric and strategies employed by journalists provided them with the means they required.

The CPI met with mixed success, as Creel maintained that the agency abided by the highest standards of journalistic objectivity, remaining "educational and informative throughout." It employed a team of writers and journalists, such as Arthur Bullard and Will Irwin, to project objectively what Creel dubbed in rather revealing terms as the "the Gospel of Americanism."[9] Creel encouraged his agency to take advantage of a variety of media, including the newly emerging forms of film and radio. Still, the CPI trafficked its message most heavily through newsprint. Creel created a government-operated newspaper wire service, Compub, which selectively supplied stories to foreign subscribers while exercising tight controls over American papers as well. Privately employed war correspondents from the United States swore oaths to "convey the truth" to the American people, found their work subjected to censorship, and were obliged to wear a green armband with a large red "C" that indicated their status as correspondents while overseas.[10]

The CPI stuck to the truth, but it utilized only those facts officials deemed as most beneficial to advancing American aims. The agency's efforts resulted in what officials at least deemed as "objective" reporting at home and abroad, which succeeded in bringing Wilson's war aims before the eyes of the world. Creel and his colleagues certainly did not minimize the success of their effort. After the war, publicist Edward L. Bernays, who served with the CPI Foreign Press Bureau and is considered the "Father of Public Relations," credited the agency with developing and disseminating Wilson's Fourteen Points.[11] Despite, or perhaps because of, its success in promoting Wilson's internationalist ideals, the CPI faced suspicion from Republicans in Congress, who made much of Wilson and Creel's friendship. The House Appropriations Committee investigated the agency during the war; Congress cut its budget in half in 1918, then abruptly terminated the agency all together in 1919 after accusing Creel of trying to monopolize coverage of the peace conference in Versailles, France.[12]

While the CPI left in its wake partisan acrimony, America's experience during World War I set a precedent for the nation's use of both information and psychological warfare during the Second World War. Despite the uproar surrounding its demise, the CPI succeeded in linking the American government's view of events with a factual and balanced perspective by its reliance on the news to convey the message and with its direction under a well-known journalist who insisted that his agency remain anchored to a factual foundation. According to Creel, the CPI sought to "inform, for such was the justice of America's case, the wonder of America's achievement, that we felt that information was in itself the most conclusive argument."[13] In addition, the CPI refused to get involved in combat propaganda despite repeated appeals made by Blankenhorn to Creel throughout the war to conduct some form of psychological warfare.[14]

NOTHING BUT THE TRUTH

Creel's efforts to maintain a distinction between propaganda and information proved an example United States President Franklin Delano Roosevelt would follow after the Japanese attack on Pearl Harbor that brought America once again into a world war in December 1941. Within months of declaring war on the Axis powers of Nazi Germany, Japan, and Italy, Roosevelt created the Office of Wartime Information (OWI) and appointed the renowned former *New York Times* reporter Elmer Davis as its director.[15] Davis possessed no experience that would make him a likely candidate to head up an agency that received a forty million dollar budget in its first year of existence, but Roosevelt made his selection fully aware of the partisan bitterness that brought about the end of Creel's CPI. Roosevelt wanted Davis to head the new agency for his "American-ness," not for his organizational skills. He believed in the stewardship Davis could provide it because his rapport with the American public would instantly earn the OWI a level of trust never achieved by the CPI.[16]

Like the CPI before it, the OWI fell primarily under the direction of journalists such as Davis and Edward W. Barrett, a former associate editor of *Newsweek* who presided over the Overseas Branch. Through the radio waves transmitted by the Voice of America (VOA) and through newspapers it distributed, such as *L'Amérique en Guerre*, the OWI resurrected Creel's "factual" approach and adhered to the notion that maintaining an objective tone in its presentation would provide the American war effort the best publicity. This "Strategy of Truth," according to Milton Stover Eisenhower, a former Director of Information for the United States Department of Agriculture who became OWI deputy director in 1942, provided Americans with the "means of describing democracy and freedom, our war aims, and our determination to win both war and the peace."[17] Under Davis's direction, the VOA promised its listeners, "Daily, at this time, we shall speak to you about America and the war—the news may be good or bad—we shall tell you the truth."[18]

The "Strategy of Truth" pursued by the OWI throughout its promotion of the American war effort during World War II carried on the tradition of the CPI of equating United States policy as closely as possible with an apparently objective perspective. Like its predecessor, the OWI relied on using only the "facts," even if it did so in a selective manner. In testimony he gave before a

Congressional committee in 1944, Davis clarified the thinking behind the OWI approach by stating, "We do not tell the truth just for fun, we use the truth in such a way as to win and shorten the war." He continued, "We tell a true story to every area; but to each one we tell the kind of true story that will best serve our interests."[19] Still, no matter how flexible they considered their policy of objectivity, OWI officials discerned a distinct difference between the activities they engaged in and those that propagandists carried out. That very perception is what prompted Roosevelt's creation of the OWI and its more unabashedly propagandistic counterpart, the Office of Strategic Services (OSS), once war had broken out.

Following the First World War, the United States possessed no apparatus for carrying out either information activities or psychological warfare. While the CPI had been dismantled by Congress, Blankenhorn's G-2-D experienced the same fate. At the conflict's conclusion, the United States Army demobilized it, turned over the subsection's remnants to the Political Section of Military Intelligence, and left them to languish there. It turns out that G-2-D made a greater impression on its adversaries than it did on those whose actions it supported in combat. During the war, the German High Command credited Blankenhorn's leaflets with having an adverse impact on the morale of its troops. The United States Army, though, resistant to the "unconventional" tactics used by G-2-D while the conflict raged, possessed little interest in further exploring the potential of a unit it considered primarily as a drain on valuable resources. By the eve of World War II, the institutional memory of G-2-D had largely disappeared from Army commands, staff schools, manuals, and doctrine. The Army did not assign a single officer to study the utility of psychological warfare in combat from 1925 to 1935.[20]

William J. ("Wild Bill") Donovan is the individual most responsible for America's resurrection of programs to conduct PSYOP.[21] He came to the Roosevelt administration as an attorney, but his military roots extended deep. In 1911, he helped organize Troop I of the First New York Cavalry of the National Guard. He earned considerable distinction during World War I by commanding the First Battalion of the 165th Regiment of the Regular Army and attaining the rank of colonel. After the war, Donovan established a law practice in New York and dabbled unsuccessfully in politics. By 1940, despite his reputation as a prominent Republican attorney and harsh New Deal critic, President Roosevelt sought Donovan's assistance in assessing the increasingly troublesome climate overseas that had resulted from the aggressive actions taken by the Axis powers. After conferring with his new secretary of the navy, Frank Knox, Roosevelt invited Donovan to the White House in July 1940. Knox, an old friend of Donovan's, recommended "Wild Bill" to the president for a special mission abroad to survey the unfolding events.

Donovan accepted Roosevelt's invitation, and over the next nine months he embarked on two separate fact-finding missions. He left for England shortly after his meeting with Roosevelt in July and returned to the United States in August before setting off once again, this time for a tour of the Mediterranean and of the Middle East. Donovan met with many foreign officials overseas and discussed a variety of subjects.[22] He came away from the experience, though, most impressed with the potential of PSYOP. During the month of Donovan's first trip to England, the British created the Special Operations Executive (SOE) to devise unorthodox methods of warfare in anticipation of a German invasion. Donovan received an appraisal of the new operation from William Stephenson, head of the British Security Coordination (BSC), the center for an intercontinental intelligence network that the British established in New York during the spring of 1940.[23]

Shortly after Donovan's arrival back in the United States in March 1941, he began preaching to anybody who would listen of the need for a rejuvenated American PSYOP program. He warned that the "Nazi military machine at work, is used not for fighting but for intimidation, to impose upon weaker countries Nazi economic and political philosophy." He attributed Nazi military successes to the regime's ability to erode the will to resist of other nations by first deploying a massive psychological offensive. Donovan observed the Nazi strategy "was just as it used to be in our schoolbooks, where we read that the soldiers of ancient days prepared for the taking of a city by first undermining its walls." He concluded, "Modern Nazi warfare is carried out in the same way."[24]

Donovan chastised America's "inability to respond effectively in the modern war of words." He called it "appallingly costly in blood and treasure and time to rely entirely, or even primarily, on orthodox methods of warfare."[25] Roosevelt listened to Donovan's advice and appointed him to direct the newly created Office of the Coordinator of Information (COI) on July 11, 1941. Donovan received the authority from Roosevelt to oversee the nation's PSYOP. Sensitive to the criticism Wilson had received for creating the CPI, Roosevelt attempted to minimize any potential backlash by limiting the new agency's jurisdiction and by publicizing its restrictions. He made it known that the COI "is not intended to supersede or to duplicate, or to involve any direction of or interference with the activities of the General Staff, the regular intelligence services, the Federal Bureau of Investigation, or other existing departments."[26]

Donovan proceeded quickly to build the COI from the ground up in a way that closely reflected four principles he considered essential for the United States to wage successful psychological warfare: white propaganda (materials such as leaflets, attributable to the United States), black propaganda (rumor campaigns and other clandestine activities), sabotage, and the synthesis of strategic information. He considered propaganda, intelligence gathering, and covert operations as proper subsets for a psychological warfare program. He considered these elements inseparable and created four integrated units within the COI to carry them out.

Donovan called his first unit the Special Intelligence Service (SIS) and charged it with obtaining "information not ordinarily obtained by other agencies of the government." Donovan's second unit, the Research and Analysis Branch (RA), consisted of a team of experts who worked to interpret the findings of the SIS and provide synthesized information to other government branches. The Foreign Information Service (FIS), Donovan's third unit, provided COI with its ability to disseminate messages. The FIS took "the fight to the enemy" by distributing American propaganda by "radio, pamphlet, leaflet, posters, advertising, and other means." Donovan established a final unit, the Special Operations Services (SOS), to carry out subversive warfare in other enemy countries. Donovan underscored that the tightly integrated units of the COI provided it with the ability to obtain "an economy of force," a condition he considered as a "vital principle in warfare."[27]

From the onset, Donovan worked to establish a close relationship between the COI and America's military. He held a "strong feeling that the preparation for any plan involving political and subversive warfare must heavily involve those entrusted with the prosecution of subsequent forms of warfare."[28] Donovan wrote to his old friend Secretary of the Navy Frank Knox about his idea to create an American counterpart to the modern British commandos and place them under COI direction. Donovan made it clear he wanted a force that stood "free from all conventions which surround a traditional corps." He confided to Knox that American success in an impending war hinged on the nation's deployment of "men calculatingly reckless with disciplined daring, who are trained for aggressive action." Donovan believed the leaders of America's conventional military services had become overly preoccupied with strategies of defense. With all of America's "talk on defense," Donovan warned Knox, the nation is "apt to miss the spirit of the attack." Donovan wanted his own battalion of psychological warriors that he could ingrain with the mentality to "always attack" and "never defend."[29]

By the spring of 1942, Donovan felt satisfied that the COI had achieved the relationship with the military that he desired it to attain. He wrote Roosevelt that the armed services "now seem to have confidence in our organization and feel that we have in motion certain instrumentalities of war useful to them."[30] At the same time, though, Donovan's efforts to align the nation's PSYOP wholly with the military to fulfill his notion that "foreign propaganda must be employed as a weapon of war" created considerable consternation among the journalists and writers who comprised the FIS.[31] Robert Sherwood, FIS chief and former Roosevelt speechwriter, labeled Donovan's ideas "reactionary." He fumed that "Bill has sold out to the Generals" and characterized Donovan's conception of aggressive psychological warfare as unnecessary and potentially counterproductive.[32]

Sherwood contended that the FIS could help the United States win the war by utilizing the same information strategy that Creel employed with the CPI during World War I. But Donovan wanted to use the FIS as a source of misinformation to obscure covert operations that the COI carried out. Donovan wrote to Roosevelt, "As long as they talk about us in the radio field, they will be unaware of what we are doing in the other."[33] Sherwood countered by insisting on America's need to use "rational methods" to convince the people of the Axis nations of the hopelessness of their cause. Donovan responded by suggesting, "Soldiers will fight this war, not the poets, and the place to help fight it is with the soldiers."[34] The intensity of the dispute left the COI paralyzed.

In a discussion that policy makers held regarding how to handle the situation, United States Under Secretary of State Sumner Welles referred to the whole matter as a "headache."[35] In this case, Welles's comments reflected the frustration caused by Sherwood and Donovan's dispute. Still, they are also emblematic of the friction that has historically plagued the relationship between America's State Department and its information program. While United States diplomats have accused information officials of attempting to circumvent their department and make policy independently, information officials have repeatedly cited the failure of State personnel to provide their agencies with timely policy guidance.[36]

Roosevelt finally intervened to quell the turmoil by dismembering the COI on June 13, 1942. He removed the FIS and placed it in the newly created OWI. He assigned what remained of the COI to the jurisdiction of the Joint Chiefs of Staff (JCS), renamed it the Office of Strategic Services (OSS), and assigned Donovan as the director. While Donovan lamented his loss of control over the FIS, he took solace in his agency's new proximity to the military. He called Roosevelt's reorganization a "sensible and necessary step toward the most effective use of all modern war weapons."[37] The OSS went to work conducting PSYOP according to Donovan's conception of them as "the arrow of initial penetration in conditioning and preparing the people and the territory in which invasion is contemplated."[38]

Through its Morale Operations Branch (MO), the OSS concentrated on spreading rumors that appeared to originate from enemy sources and could create a climate of confusion and disorganization. In addition, it conducted clandestine radio broadcasts and dispersed leaflets aimed at undermining enemy morale. These materials, in part, resulted from the efforts of one Heber Blankenhorn, who Donovan hired in early 1943 and assigned to MO based on his World War I experiences.[39]

BATS AND FOXES, OH MY!

Under the direction provided to them by Davis and Donovan, the OWI and the OSS went their separate ways carrying out PSYOP during World War II. Despite Donovan's unconventional ideas, his organization fared far better than the OWI did at gaining the respect of American officials during the course of the war. While Donovan's OSS made a particular impression on United States military leaders, Davis spent much of his time defending the OWI against the criticism leveled at it for much of its existence.[40] Staffed by New Deal liberals such as Sherwood and the pollster Elmo Roper, OWI utilized its overseas radio station, the Voice of America (VOA), to criticize wartime decisions made by the United States for the sake of expediency that appeared to compromise liberalist ideals that promoted the victory of democracy over totalitarianism as America's principal war aim.

Most famously, New Dealers questioned American decisions to support governments in Italy and North Africa in order to facilitate the war against Nazi Germany. Following the Allied invasion of French North Africa in November 1942, for example, United States Army General Dwight David Eisenhower chose to neutralize French military resistance by acknowledging the sovereignty over the region claimed by Marshal Pétain's Vichy-based government of unoccupied France. This decision meant excluding the French forces led by General Charles de Gaulle, which had refused to collaborate with the Nazis and the Vichy government. OWI New Dealers alleged that alliances such as this one, however sensible from a military perspective, sent the wrong message to those resisting the Nazis elsewhere. These pacts made for the sake of military expediency, they contended, risked leaving conquered peoples in Europe wondering "if a precedent had been set for dealing with Hitler's stooges after liberation."[41]

In the case of the French in North Africa, the OWI prevailed on President Roosevelt to explain the decision as "a temporary expedient justified solely by the stress of battle." A year later, OWI expressed its disdain for American decision-making publicly. Following the Allied invasion of Italy, the United States recognized a new government named by King Victor Emmanuel III, who OWI New Dealers identified as a confirmed fascist because of his previous toleration of the dictator Benito Mussolini. One VOA program went as far as to call Italy's King Victor Emmanuel III a "moronic little king." It further suggested that the Allies should not celebrate prematurely because the Italian fascists continued to fight in the war.

In addition to the tension between American officials and their information agency regarding Italy and Africa, pamphlets the OWI produced that touted postwar self-determination for colonial peoples as an American war aim routinely failed inspection from the State Department. Foreign Service professionals believed in elite diplomacy and viewed the OWI as a potentially disruptive force in America's foreign policies, particularly when it promised postwar liberation to people colonized by European allies of the United States. The State Department rejected one OWI initiative that would have advocated self-determination in India, for example, for fear it might incite an Indian insurrection against the British.[42] By the end of the war, the OWI faced the same predicament the CPI had encountered twenty years earlier, dissolution by a Congress that viewed its activities as more partisan than informational.

The wartime experience of the OSS proved different from its OWI counterpart. Under Donovan's direction, the agency steered clear of the war's political issues and aligned itself closely with America's military, a situation made easier by Roosevelt's decision to put the OSS under the authority of the Joint Chiefs. Still, indicative of the American military's conception that combat propaganda possessed only limited utility, the JCS did not issue the OSS its first definitive charter (JCS 155/4/D) until December 23, 1942, and it only approved of activities carried out by the MO branch belatedly by granting it official status (JCS 155/11/D) on October 27, 1943, ten months after Donovan had created it.

The evidence suggests that by the end of the war, though, the OSS had achieved a great measure of appreciation for its activities from American officials. Particularly within America's military, leaders who had initially expressed their skepticism regarding the value of PSYOP had become convinced of their usefulness in undermining enemy morale. With the charter it finally provided MO, the JCS enabled the OSS to conduct clandestine operations "within enemy countries, and enemy-occupied or controlled countries, and from bases within other areas, including neutral areas." The charter empowered Donovan's agency to execute "forms of moral subversion of diverse means" and to utilize "false rumors, 'freedom stations,' false leaflets and false documents, for undermining the morale of the enemy."[43] For Donovan, it provided the endorsement he sought for his idea that PSYOP served as the modern day equivalent of when "the soldiers of ancient days prepared for the taking of a city by first undermining its walls."[44]

By the time American officials turned their attention toward a possible invasion of Japan in 1945, Donovan's boasts about OSS success sparked little protest. Donovan bragged that his agency had "gained experience and developed operating techniques in a field never developed by any U.S. agency prior to this war."[45] United States Army Chief of Staff General of the Army George C. Marshall, a skeptic of Donovan's agency throughout much of the conflict, had come to concur with this assessment by the summer of 1945. Marshall suggested that a "full exploitation of psychological warfare may bring about favorable results" in Japan, "both in the matter of shortening the war and inducing an increased number of battlefield surrenders."[46] No doubt important to the high regard officials came to hold of the OSS is the advocacy of it provided by United States Army General Dwight David Eisenhower. Eisenhower overcame his "soldierly distrust" of PSYOP quickly, watching Donovan's MO Branch spread rumors he deemed as effective throughout first North Africa and then Europe.[47] By war's end, Eisenhower believed that "Psychological warfare has proved its right to a place of dignity in our military arsenal."[48]

Eisenhower's belief in the potential of the propaganda component of PSYOP, while important for OSS support during the Second World War, is additionally noteworthy, as it proved essential in shaping the postwar program that Americans constructed in Vietnam. Eisenhower had actually become aware of the importance of hearts and minds during his posting in the Philippines in the 1930s, where he engaged in nation-state building. The experience taught him that security programs could only succeed if first the heartfelt allegiance of the indigenous population could be secured.[49] This lesson proved to be an important one for Eisenhower, as it not only opened him to the possibilities afforded by psychological warfare during World War II, but it also influenced his thinking regarding the Cold War.

Following the Nazi surrender in 1945, Eisenhower became the military governor of occupied Germany, a position that placed him on the front line of the growing Soviet-American hostility. By the end of the year, he accepted an appointment to become the United States Army chief of staff, from which he helped craft the provisions of the 1947 National Security Act. This law intended to bring about improved coordination within America's armed forces and created the Department of Defense to house the Departments of the Army, Navy, and Air Force. Eisenhower left his Army post in 1948 to become the president of Columbia University, but his retirement from the military did not last long. By the end of a year in which the Soviet Union blockaded Berlin and a communist coup transpired in Czechoslovakia, Eisenhower accepted Secretary of Defense James Forrestal's request to temporarily chair the Joint Chiefs of Staff.

By the middle of 1950, the Soviets possessed the atomic bomb, the communist party under Mao Zedong ruled China and war erupted in Korea. President Harry S. Truman, concerned about the appearance of European vulnerability because of the developments in Asia, prevailed on Eisenhower to accept the position of the first supreme allied commander (SACEUR) of the newly created North Atlantic Treaty Organization's (NATO's) armed forces. According to Robert R. Bowie, Director of the Policy Planning Staff, Department of State (1953–1957), and eminent historian Richard H. Immerman, the various roles Eisenhower assumed during the early postwar years enabled him to sharpen "beliefs and skills that had been germinating over three decades."[50]

Eisenhower had grown convinced that the Communist threat America faced necessitated the nation's development of a strategic concept from the foundation of an "enlightened appreciation of what constituted national security." Bowie and Immerman successfully demonstrate that Eisenhower defined national security, due to his experiences during the war and after, as comprising a "broad range of interests, especially a healthy economy."[51] Eisenhower perceived national security and national solvency as mutually dependent, because if America's economy crumbled due to an excessively expensive peacetime defensive program, the nation's "core values" would be endangered as well. A strained economy could well necessitate the institution of un-democratic measures, such as rationing, price controls and censorship—a situation Eisenhower believed would play right into the hands of Soviet propagandists. In this sense, hearts and minds stood out to Eisenhower as crucial components of the Cold War. He believed that, by "appealing to the cause of freedom, democracy, and individualism, the free world could unmask communism for the oppressive system that it was."[52]

Elected to the presidency of the United States in 1952, Eisenhower prioritized the creation of a postwar PSYOP program that continued the traditions of the OSS through the personnel that both developed and directed it. C. D. Jackson, a vice president of Time-Life at the time of Eisenhower's election, headed the Psychological Warfare Division (PWD), which coordinated combat propaganda operations in the European theater during World War II. Eisenhower looked to Jackson to provide American PSYOP not only with the vigor demanded by the Cold War, but also with the planning and organization that had characterized the United States effort during World War II.

Eisenhower brought Jackson to the White House with him as the first special assistant to the president for Cold War planning. In his work on the USIA during the 1950s, historian Walter L. Hixson labels Jackson as the "ultimate psychological warrior of the Eisenhower team."[53] While that may have been the case, Jackson served Eisenhower in the company of others that the president surrounded himself with as well, including OSS veteran Abbott Washburn, who proved highly influential in the early years of the USIA. Washburn served on the committee chaired by William Jackson (unrelated to C. D. Jackson) that brought about the creation of the USIA in 1953, and he acted as the agency's first deputy director.

More so than for any other American president, psychological strategy for Eisenhower signified the interconnectedness of "international public opinion, persuasion, and national security policy." Rather than situating PSYOP on the margins of international diplomacy, Eisenhower enabled it to emerge as a "significant component" of American foreign relations. He overtly utilized the USIA to its fullest while simultaneously endorsing "camouflaged" activities that officials set in motion by utilizing independent news media, nongovernmental organizations, and private individuals as "surrogate communicators" for carrying "propaganda messages." Eisenhower laid the foundation for the total war approach to PSYOP that Americans would build the JUSPAO atop in Vietnam.[54]

While the individuals who Eisenhower appointed to carry out American PSYOP brought with them the OSS concept of aggressive and total propaganda Donovan had espoused, they also carried with them their former agency's proclivities for characterizing European and Asian audiences on starkly different terms. Historians have shown that the Pacific theater of World War II contained a racial dimension that did not exist in other areas of operation.[55] Until recently, this assertion went uncontested regarding American PSYOP against Japan, as, due to President Roosevelt's Europe first strategy, these activities "never really got off the ground in the Far East."[56] In her book *You Can't Fight Tanks with Bayonets*, Allison B. Gilmore studies the limited operations the United States ultimately did carry out against the Japanese in the Southwest Pacific. She contends that because Americans did proceed with combat propaganda against the Japanese Army, questions must be raised about the assertion made by scholars that racial hatred against Japan motivated United States leaders in the way they waged war in the Pacific. She writes:

> For the Allies to have waged psychological warfare only against Germans and Italians without making an equivalent attempt to persuade Japanese to give up the fight would lend credence to claims that in the Pacific War the enemy was regarded as subhuman, beyond rationality, and worthy only of a kill-or-be-killed approach. But that was not the case.[57]

Gilmore's case is a solid one, illustrating that Americans did not consider the racial extermination of the Japanese as their wartime priority. Nevertheless, a decidedly racialist perception did permeate the minds of American planners in the OSS as they turned their attention from Europe to Asia. While there is no indication these individuals considered the Japanese subhuman, the evidence does suggest they considered Japanese as decidedly more primitive than their other adversaries in Europe. In his analysis of Allied PSYOP during the Second World War, Daniel Lerner, a sociologist who served as an intelligence officer in the United States Army during the war, demonstrates that campaigns carried out against Nazi Germany, whether overt or covert, "were obliged to observe the rule of accurate reporting."[58] Americans believed that in order for PSYOP to be effective in Europe, they needed above all else to be credible by containing at least a kernel of truth.

Historian Clayton D. Laurie supports Lerner's observation by illustrating the prohibition officials ordered against spreading atrocity stories about "human sausage, human skin decorations, and canned human meat" because they deemed these tales to exceed all reasonable boundaries of belief.[59] PSYOP in Europe focused appeals on the audience's intellect in an effort to plant suggestions in its member's minds of a lost cause. The aim remained throughout the war to eradicate the will of America's European foes by playing on their rational concerns about death, loss, and betrayal. Chiefly through rumors that it generated, the OSS sought to accomplish tasks such as raising questions in German minds regarding the motives of their Nazi leaders and building on Italian suspicions of Germany.

The aims of PSYOP against Japan proved quite different from those employed in Europe. From the beginning, American planners considered the Japanese as more susceptible to primal appeals than Europeans and looked to implement tactics of terror "that strike past the reason of the victim and into the irrational."[60] They received an early push in this direction from President Roosevelt, who approved an unusual plan that a private citizen, Lytle S. Adams, sent to him from Pennsylvania. Adams recommended to Roosevelt that the United States should launch a surprise attack against Japan by strapping "incendiaries around the necks of bats" and flying them into Japan.[61]

Adams contended that, besides the damage that the explosives would do, the bats alone would terrorize the Japanese population, who collectively held a mortal aversion to the winged creatures. Roosevelt, who apparently found the idea plausible, sent it to Donovan on January 12, 1942 for him to develop further. Donovan pushed the MO Branch to make the plan operational and recruited Dr. Remington Kellogg, the American Museum of Natural History's curator of the division of mammals along with the United States Army Air Corps to work out the plan's finer points.

By May 1943, Donovan remained optimistic about the ability of the OSS to set its bats loose over Japan. He wrote that Adams's "idea is being thoroughly checked," and "so far it appears to have met all tests."[62] In the end, however, insurmountable obstacles prevented the OSS from bringing the plan to fruition. It turned out that bats do not travel well in high altitude transport planes. The cold air induced the bats into a "stupor" that tended to result in them plummeting into the sea when released by their handlers. In addition, planners never managed to figure out how to teach the bats "to fly to Japan and not return to their vessels."[63] The OSS had to abandon its bat plan, but as it turns out, the scheming conducted by the agency's officials on how best to instill their Asian enemies with supernaturally inspired terror had only begun.

Along with their British allies, Americans perceived a "weakness in the psychology of the Japanese" that they could not meet "by ordinary Western means" nor measure "solely by Western standards."[64] They strove to "go deeply into beliefs of primitive origin which are currently held by modern Japanese."[65] They considered this tactic a necessity because, along with their British allies, they held that the Japanese "do not attach the same values to life and death" that apply in the West. Planners saw irrationality as the hallmark of Japan's culture and believed appeals to reason would be of little use. The operations they devised strove to drive Japan into a state of "general despondency and even terror."[66] To do so, officials compiled a list of characteristics they alleged to typify Japanese that included, "super-aggressiveness, face-saving, tradition of self-destruction, inflexibility, and brutality."[67]

Americans observed few of the boundaries that they had constructed to constrain their output in Europe. As they anticipated a United States invasion of Japan's home islands, they planned to tap into the wild emotions and irrational urges they believed held sway over Japanese actions and could lead to a quick collapse of the empire. Planners demonstrated no hesitancy to devise "certain tactics of which we might otherwise disapprove but that are justified when used against the Japanese."[68] They identified themes to exploit from data they collected from "persons who know the Japanese" and from their own "extensive research in anthropological literature, travel literature, and comparative relations." They focused on creating a widespread sense of terror in Japan by utilizing the "superstitions currently prevalent among all classes of Japanese" and proposed such tactics as poisoning Japan's "sacred trees" and painting some of its "sacred white animals" on America's armored vehicles.[69]

When it came to developing PSYOP against the Japanese, the OSS proved willing to consider anything that its planners thought might ridicule Japan's high command, its upper classes, and even its emperor. Agency officials entertained the idea of dropping balloons over Japan that depicted the emperor with the left side of his head shaven to mock the beliefs that the Japanese held in him. They also sought to exploit the view that many Americans maintained about the prevalence of homosexuality among Japan's troops. They intended to develop a campaign that might "excite the nostalgic yearning" of Japanese soldiers for "geishas and the flesh pots of Yoshiwara."[70]

The core of OSS planning contained plans to use the fox, an animal that, when illuminated, Shintoists view as a harbinger of bad times. "Project Fantasia" stemmed from the belief that "under extremely trying conditions," Japanese "would be adversely affected by what they might consider an evil omen." Based on what they learned from "experts familiar with Japanese superstition and mythology," OSS personnel hoped the project would instill their enemy with "fear, terror, and despair."[71] Beginning in 1943, Americans experimented with fox-shaped balloons covered in luminous paint that could be dangled by fishing line to scare Japanese observers. By the end of 1944, they shelved that idea and shifted to testing live foxes sprayed with paint they resolved to disperse across the "entire field of combat." They recognized the potential of the fox legend if turned "toward the morale" of Japan's civilian population and thought of their plan as America's "most potent" psychological tool against the Japanese.[72]

According to the final plans, "Operation Fantasia" intended to commence with the distribution of pamphlets warning of impending evil and patterned after those used by Japanese soothsayers. This material would be dropped by air and also distributed by operatives in the field who carried specially constructed reed whistles that simulated a fox-like "cry of the damned." The OSS had contracted the chief chemist at the United States Radium Company to reproduce "fox odors" in the form of both powder and paste for those individuals in the field who handed out the pamphlets. As a final touch, "Operation Fantasia" planners called for the use of collaborators from among the Japanese population to "simulate persons possessed of the Fox spirit."[73] To the chagrin of New Yorkers, the OSS tested its idea during the height of the war by releasing thirty foxes in Central Park that "were painted with a radiant chemical which glowed in the dark." As a result, "Horrified citizens, shocked by the sudden sight of the leaping ghostlike animals, fled from the dark recesses of the park with the 'screaming jeemies.'"[74]

Satisfied with their trials, American planners set about procuring all of the foxes they could get their hands on from China and Australia in anticipation of their nation's impending invasion of Japan. Only the war's sudden conclusion, brought about by America's use of the atomic bomb on Hiroshima and Nagasaki, denied them the opportunity to inundate Japan with their glowing foxes. Still, the development of their idea demonstrates how Americans during the war perceived the psychology of their Asian foe in a far different way than they saw their enemies in Europe. Based on their notions of Japan's primitive state, Americans produced plans like "Operation Fantasia" for use against Japanese that stood as much more absurd than any European campaign they proved willing to consider.

Even the horrified reaction of New Yorkers suddenly trapped in the atmosphere of a bad B-movie, likely not a Shinto among them, did not shake officials from their conviction that Asians maintained a monopoly in the department of irrational superstitious beliefs. Neither would their encounters with other Asians in the years that followed Japan's defeat, when Americans would continue to ground their psychological appeals in the foundation provided by anthropologists and others whose expertise they deemed sufficient to enable them to penetrate the mysterious oriental psyche they sought repeatedly to reach with their combat propaganda.

Americans left their foxes at home whenever it came time for them to depart for Vietnam. They did, however, resume their targeting of a "superstitious enemy."[75] In doing so, their primary aim remained consistent with the PSYOP tradition of the OSS, to induce defections and surrender through a campaign run jointly with the Vietnamese Information Service (VIS) called *Chieu Hoi* (Open Arms). Ninety-five percent of the leaflets generated by American PSYOP consisted of surrender appeals in support of this program.[76] In order to make it a success, officials found it necessary to devise "guidelines for the exploitation of enemy vulnerabilities provided by superstitions and deeply-held traditional beliefs." They endorsed the use of hooting owls and calling crows to target "rural" Vietnamese who Americans believed could be "provoked into a fear response if startled at night" by these "death omens."[77]

The extent to which Americans were convinced that Vietnamese stood as susceptible to manipulation through irrational appeals is evident through the "Ace of Spades" leaflet. Despite intelligence that demonstrated that this symbol had no meaning for Vietnamese as the "symbol of death" that Americans had assumed it did, the leaflet continued to be circulated throughout the course of the entire war.[78]

More than any other type of appeal, American leaflet and loudspeaker messages harped on the theme of death. One leaflet used asserts, "Your death is only a matter of time." On its reverse side, the leaflet displays a photograph of a dead soldier.[79] Others depict photos and sketches of decapitated bodies, mass graves and a variety of skulls, some roasting in flames and others emblazoned with the ace of spades, a sign erroneously assumed by Americans to represent an evil omen for Vietnamese. Still more of these leaflets show grotesquely twisted corpses suffering from the advanced stages of rigor mortis. According to former JUSPAO officer Robert Chandler, death themes "were repeated over and over in virtually all enemy-oriented communications." Here, the style of PSYOP shaped by the American Way of War is evident. Death themes proved most popular despite intelligence, according to

Chandler, that leaflets themed by "brotherly love" and "happiness of family" proved much more likely to induce defection than did the "rally or die" motif preferred by American PSYOP.[80]

NOTES

1. Don Luce, "Tell Your Friends that We're People," in Noam Chomsky and Howard Zinn, eds., *The Pentagon Papers: The Defense Department History of United States Decision-Making in Vietnam*, Gravel edition (Boston: Beacon Press, 1972), V: 95.

2. Phan Van Hoang (People's Liberation Army), in *Cultures in Conflict: The Vietnam War*, ed. Robert E. Vadas (Westport, Connecticut, and London: Greenwood Press, 2002), 141.

3. Harry S. Truman, "Today's Need for A Campaign of Truth," in *America's Weapons*, 31.

4. Zorthian, "The Use of Psychological Operations," 40.

5. Bruce Bidwell, *History of the Military Intelligence Division of the War Department General Staff: 1775–1941* (Frederick, Maryland: University Publications of America, 1986); George G. Buntz, *Allied Propaganda and the Collapse of the German Empire in 1918* (Stanford, California: Stanford University Press, 1938); and Clayton D. Laurie, "'The Chanting of Crusaders': Captain Heber Blankenhorn and AEF [American Expeditionary Forces] Combat Propaganda in World War I," *The Journal of Military History* 59, no. 3 (July 1995): 457.

6. George Creel, *How We Advertised America* (New York: Harper & Brothers Publishers, 1920); James R. Mock and Cedric Larson, *Words that Won the War: The Story of the Committee on Public Information, 1917–1919* (New York: Russell and Russell, 1939; reprint, 1968); and Stephen L. Vaughn, *Holding Fast the Inner Lines: Democracy, Nationalism and the Committee on Public Information* (Chapel Hill: University of North Carolina Press, 1980).

7. Creel, *How We Advertised America*, 9.

8. M. L. Sanders, "Wellington House and British Propaganda During the First World War," *Historical Journal* 18, no. 1 (1975): 122; and Philip M. Taylor, "The Foreign Office and British Propaganda During the First World War," *Historical Journal* 23, no. 4 (December 1980): 876.

9. Rosenberg, *Spreading the American Dream*, 80.

10. Knightley, *The First Casualty*, 122.

11. Robert F. Delaney, "Psychological Operations in the 1970s: A Program in Search of a Doctrine," in *The Art and Science of Psychological Operations*, I: 1.

12. Sorensen, *The Word War*, 8.

13. Creel, *How We Advertised America*, 9.

14. Laurie, "The Chanting of Crusaders," 461.

15. Edward W. Barrett, *Truth Is Our Weapon* (New York: Funk & Wagnalls, 1953); Roger Burlingame, *Don't Let Them Scare You: The Life and Times of Elmer Davis* (Philadelphia: J. B. Lippincott Co., 1961); Wallace Carroll, *Persuade or Perish*

(Boston: Houghton Mifflin, 1948); Milton S. Eisenhower, *The President Is Calling* (Garden City, New Jersey: Doubleday & Company, Inc., 1974); Robert William Pirsein, *The Voice of America: An History of the International Broadcasting Activities of the United States Government, 1940–1962* (New York: Arno Press, 1979); Holly Cowan Shulman, *The Voice of America: Propaganda and Democracy, 1941–1945* (Madison: The University of Wisconsin Press, 1990); Allan M. Winkler, *The Politics of Propaganda: The Office of War Information, 1942–1945* (New Haven and London: Yale University Press, 1978); and Sydney S. Weinberg, "What to Tell America: The Writer's Quarrel in the Office of War Information," *Journal of American History* 55, no. 1 (June 1968): 73.

16. Burlingame, *Don't Let Them Scare You*, 47.

17. Eisenhower, *The President*, 128.

18. Pirsein, *The Voice of America*, 58.

19. Statement by Elmer Davis, director, Office of War Information, U.S. Congress, House, Committee on Appropriations, *National War Agencies Appropriations Bill, Hearings before a Subcommittee of the House Committee on Appropriations,* 78th Cong., 1st sess., 1944, 707.

20. Laurie, "The Chanting Crusaders," 474.

21. Anthony Cave Brown, *The Last Hero: Wild Bill Donovan* (New York: Times Books, 1982); Richard Dunlop, *Donovan: America's Master Spy* (New York: Rand McNally, 1982); Corey Ford, *Donovan of OSS* (Boston: Little, Brown and Company, 1970); and Thomas F. Troy, *Donovan and the CIA: A History of the Establishment of the Central Intelligence Agency* (Frederick, Maryland: Aletheia Books, University Publications of America, Inc., 1981).

22. Troy, *Donovan and the CIA*, 28.

23. William Stevenson, *A Man Called Intrepid: The Secret War* (New York and London: Harcourt Brace Jovanovich, 1976); and Thomas F. Troy, *Wild Bill and Intrepid: Donovan, Stephenson, and the the Origin of the CIA* (New Haven and London: Yale University Press, 1996).

24. "Ships Must Get Goods to Britain to Assure our Aid, says Donovan," *New York Times,* March 27, 1941, 1.

25. Wallace Deuel, *History of OSS* (5 volumes, undated, unpublished), I: 99, box 122A, William J. Donovan Papers, MHI.

26. "Donovan is Named Information Head," *The New York Times,* July 12, 1941, 1.

27. Letter, Donovan to Smith, March 23, 1942, *Records of the JCS,* XI: F248.

28. Deuel, *History of OSS*, II: 26.

29. Letter, William J. Donovan to Secretary of the Navy Frank Knox, July 3, 1941, box 119A, Folder 22, Donovan Papers, MHI.

30. Memo, Donovan to Roosevelt, June 30, 1942, box 119A, Folder 22, Donovan Papers, MHI.

31. Kermit Roosevelt, *War Report of the OSS* (2 volumes, New York: Walker and Company, 1976), I: 223.

32. Deuel, *History of OSS*, I: 2.

33. Deuel, *History of OSS*, I: 19.

34. Deuel, *History of OSS*, I: 2.

35. Meeting of the Liaison Committee, June 3, 1942, *Records of the JCS*, XI: F253.

36. Reuben S. Nathan, "Making Policy is not the Propagandist's Business—Or Is It?" in *The Art and Science of Psychological Operations*, I: 125.

37. Memo, Donovan to Roosevelt, June 30, 1942.

38. Letter, Donovan to Smith, March 23, 1942, *Records of the JCS*, XI: F246.

39. Laurie, "The Chanting Crusaders," 275.

40. Hixson, *Parting the Curtain*; Sorensen, *The Word War*; Shulman, *The Voice of America*; James P. Warburg, "'The Moronic Little King Incident'"; and Allan M. Winkler, *The Politics of Propaganda.*

41. Sorensen, *The Word War*, 13.

42. Hixson, *Parting the Curtain*, 2.

43. Psychological Warfare Against Japan, Report by the Joint Staff Planners, July 26, 1945, *Records of the JCS, Part I: 1942–1945, The Pacific Theater*, VII: F222.

44. "Wild Bill's Brain Bureau Set Up," *New York Times*, September 21, 1941, 1.

45. Establishment of Intelligence Network and Subversive Activities in Japan, Memorandum for the Joint Chiefs of Staff from Director of Strategic services, July 18, 1945, *Records of the JCS*, VIII: F342.

46. Psychological Warfare Against Japan, Memorandum by the Chief of Staff, U.S. Army, June 17, 1945, *Records of the JCS*, VII: F218.

47. Carroll, *Persuade or Perish*.

48. Sorensen, *The Word War*, 20.

49. Robert R. Bowie and Richard H. Immerman, *Waging Peace: How Eisenhower Shaped an Enduring Cold War Strategy* (Oxford: Oxford University Press, 1998), 52.

50. Bowie and Immerman, *Waging Peace*, 43.

51. Bowie and Immerman, *Waging Peace*, 44.

52. Bowie and Immerman, *Waging Peace*, 52.

53. Hixson, *Parting the Curtain*, 22.

54. See Kenneth Osgood, *Total Cold War: Eisenhower's Secret Propaganda Battle at Home and Abroad* (Lawrence, Kansas: University Press of Kansas, 2006).

55. See Dower, *War Without Mercy*; and Fyne, *The Hollywood Propaganda.*

56. Roosevelt, *War Report of the OSS*, xiv.

57. Gilmore, *You Can't Fight Tanks*, 147.

58. Daniel Lerner, *Psychological Warfare Against Nazi Germany: The Sykewar Campaign, D-Day to VE-Day* (Cambridge: The M.I.T. Press, 1971), 28.

59. Clayton D. Laurie, "Black Games, Subversion, and Dirty Tricks: The OSS Morale Operations Branch in Europe, 1943–1945," *Prologue* 25, no. 3 (fall 1993): 265.

60. Deuel, *History of OSS*, I: 2.

61. Interview with Dr. Remington Kellogg, June 14, 1945, box 119A, Folder 25, Donovan Papers, MHI.

62. Letter, Donovan to McIntyre, May 28, 1943, box 119A, Folder 25, Donovan Papers, MHI.

63. Interview with Dr. Remington Kellogg, June 14, 1945.

64. Report, Psychological Warfare Proposals Against the Japanese, March 13, 1945, Folder 25, box 119A, Donovan Papers, MHI.

65. Report, Psychological Warfare Proposals, March 19, 1945.

66. Anglo-American Outline Plan for Psychological Warfare Against Japan, May 21, 1944, *Records of the JCS) Part I: 1942–1945, The Pacific Theater,* VIII: F190–92.

67. Operational Plans for OSS Activities in Pacific Ocean Areas, December 14, 1944, *Records of the JCS*, VII: *F267.*

68. Memo, Psychological Warfare Proposals Against the Japanese, June 21, 1943, Folder 25, box 119A, Donovan Papers, MHI.

69. Memo, Psychological Warfare Proposals, March 13, 1945.

70. Memo, to Taylor, Knapp, and Richards from E. Salinger, June 21, 1943, box 119A, Folder 25, Donovan Papers, MHI.

71. Memo, Psychological Warfare Proposals Against the Japanese, July 5, 1943, box 119A, Folder 25, Donovan Papers, MHI.

72. Memo, Psychological Warfare Proposals, March 13, 1945.

73. Memo, Psychological Warfare Proposals, March 13, 1945.

74. Article, "OSS Plan to Fox Japs with Ghost Foxes Recalled," *Sunday Tribune*, March 21, 1948, 1.

75. "Psywar," *Time* magazine, October 29, 1965, 36.

76. Volume 1, Final Technical Report Psychological Operations Studies—Vietnam, May 25, 1971, folder 01, box 15, Douglas Pike Collection: Unit 03—Insurgency Warfare, VA-TTU.

77. Policy Number 36, The Use of Superstitions in Psychological Operation in Vietnam, May 10, 1967, folder 12, box 13, Douglas Pike Collection: Unit 03—Insurgency Warfare, VA-TTU.

78. Chandler, *War of Ideas*, 141.

79. Leaflet, Finished When You Die, undated, "Vietnam Trail Leaflets," box 14, Papers of Leonard Marks, Director, USIA, LBJL.

80. Chandler, *War of Ideas*, 48.

Chapter Five

The Truth Will Set Them Free

OVERVIEW

This chapter is divided into five parts. The first, "An Instrument of Political Warfare," surveys the developments in America's PSYOP program in the immediate aftermath of the Second World War. It chronicles how neither the OSS nor the OWI survived World War II, and, while the Truman administration believed an official and purportedly "objective" program could serve America's interest in waging an ideological Cold War against the communist Soviet Union, no single agency emerged to carry out this task.

The second section of this chapter, "Come on Out, or We'll Blow Your Heads Off," shows that, with no other agency in place when the United States went to war in Korea in 1950, the United States Army assumed the role of carrying out PSYOP on behalf of the American initiative. The military effort proved consistent with its history of engagement in these types of activities, as they remained largely confined to leaflet and loudspeaker operations that leaders believed would consume minimal resources better devoted elsewhere and would have the greatest potential for having an immediate effect on the battlefield. The effort in Korea stands as noteworthy in two other regards. First, it underscored the need in the minds of American leaders in Washington for the creation of a permanent PSYOP agency. Second, it operated on the assumption that the Koreans constituted little more than Cold War pawns.

The third section examines the American effort to create a Cold War PSYOP program that brought about the United States Information Agency (USIA) in 1953. "War without Frontiers" demonstrates that, as a result of events overseas such as those in Korea and the anticommunist climate that existed at home, the USIA positioned itself as both an "objective" information agent and a tool that could be utilized against communism in a more "hard-hitting" fashion. For the first time in American history, the United States possessed an "umbrella" organization for PSYOP that dispensed both "truth" and "propaganda." By the end of the 1950s, with no end to the Cold War in sight, the "objective" side of the USIA mission became prominent in the agency's programming aimed at Europe.

"Puppeteers and Pawns," the chapter's fourth section, demonstrates that the agency continued to emphasize its propaganda mission as it began working in Asia to support United States policies, often in conjunction with individuals who had served in the OSS during World War II. Not only did the agency take a more "hard hitting" approach toward its Asian audiences it also set campaigns in motion that presumed a high level of susceptibility to outsiders to exist among people it considered to occupy a lower stage of development than did America. The USIA shared in and helped perpetuate this belief while it contented itself with publicizing the appearance of democratic reform for the sake of achieving societal stability.

This chapter's last part, "A Vietnamese Magsaysay?" conveys the efforts of the United States to transfer the PSYOP program that it ran with apparent success in the Philippines to Vietnam. The American program that began in 1954 laid the foundation for the more expanded effort that accompanied direct United States military intervention in the 1960s. It viewed Vietnamese as a people who lacked a national identity and assumed the United States could create one for them through the use of PSYOP techniques and the creation of a stable government that appeared democratic. It illustrates that this strategy deployed under the initial American team generated a backlash among Vietnamese, who viewed the government installed in Saigon as a puppet regime of the United States. The American policy provoked Vietnamese agency and national pride, a phenomena that United States officials denied as even possible in Southeast Asia.

AN INSTRUMENT OF POLITICAL WARFARE

The United States Congress dismantled the OSS and the OWI following World War II much as it had ended the Creel Committee after the First World War. America's legislative branch deemed the peacetime existence of these departments as an unwarranted threat to the nation's democratic tradition, particularly following the controversies surrounding the OWI in Italy and North Africa. In addition, Donovan's aggressive style earned the OSS a considerable number of critics at home. Chief among them stood John Edgar Hoover, the longtime director of the Federal Bureau of Investigation (FBI). Hoover resented the ability to collect and analyze intelligence in Europe and

Asia that President Roosevelt had granted the OSS. When Donovan proposed to Roosevelt in 1944 the idea of transforming the OSS into a postwar "central intelligence service," Hoover derailed the idea by leaking Donovan's plan to the anti-Roosevelt Chicago *Tribune*, which openly criticized the notion.[1]

Harry S. Truman became president following Roosevelt's death in April 1945, and World War II came to an end with Japan's surrender four months later. By the beginning of 1946, Truman recognized the inadequacy of America's fragmented intelligence apparatus and directed the establishment of the Central Intelligence Group (CIG) to coordinate, plan, evaluate, and disseminate intelligence. This mechanism proved ineffective, though, as it depended on the Departments of War, Navy, and State for both its budget and personnel. The old departments retained their autonomy over their own intelligence operations, and turf-guarding struggles ensued.[2]

The 1947 National Security Act aimed to better coordinate America's national defense and solve the bureaucratic infighting that plagued Washington. It separated the Air Force from the Army, gave the Joint Chiefs of Staff a statutory basis, created the Department of Defense, established the National Security Council (NSC) to advise the President, and rechristened the CIG as the Central Intelligence Agency (CIA). The law provided for the CIA to exist independently of the Department of Defense and to answer directly to the President through the NSC. Moreover, the legislation sanctioned the CIA to carry out covert activities in addition to collecting foreign intelligence.[3] Over the next decade, the new agency established a variety of front organizations that utilized émigrés to broadcast and to publish diatribes against the Soviet system. The CIA covertly funded and directed operations such as Radio Free Europe (RFE), which began broadcasting into Czechoslovakia, Romania, Hungary, and Poland in 1950. This radio operation conducted "a no holds barred" psychological offensive against the Soviet Union using unattributed funds.[4]

In 1948, Congress passed the Smith-Mundt Act. The legislation provided the legal foundation for the eventual creation of the USIA in 1953 by authorizing the United States government to utilize all means of communication to promote a greater understanding of American policy throughout the world.[5] In the interim, America's resurgent PSYOP program had its funding doubled to $31.2 million. As historian Walter L. Hixson proposes, the program benefited from a new American awareness of the Cold War's urgency, brought on by events overseas such as the Berlin Blockade and the establishment of the communist People's Republic of China.[6] President Truman's Campaign of Truth, which found expression over the radio waves through a resurrected Voice of America, received its direction from Edward Barrett, a *Newsweek* editor who had worked closely with Donovan in the OSS during the Second

World War. Consequently, the Campaign amounted to an OSS style offensive that aimed to undermine Soviet influence around the globe.[7]

COME ON OUT, OR WE'LL BLOW YOUR HEADS OFF

Despite the events of the late 1940s that created a consensus among Americans regarding the threat posed to their nation by communist expansion, officials took no steps to create a new instrument for PSYOP. The need to create one finally crystallized in the minds of American officials who considered their nation caught off guard when war broke out in Korea during 1950. With no existing agency able to conduct PSYOP there, the United States military took the lead. The Far East Command reactivated the Psychological Warfare Branch (PWB) and appointed J. Woodall Greene as United States Army Chief of the Psychological Warfare Branch, G-2, Far East Command.

America's military had historically demonstrated little interest in utilizing PSYOP in anything more than a tactical way. The idea took hold within the United States Army during the First World War that any value derived from psychological instruments resided in whatever was their immediate impact.[8] Such thinking continued during the Second World War. The civilian-dominated OWI and OSS carried out PSYOP in what appeared to conventionally trained soldiers as an unorthodox and undisciplined technique that intended to influence the diplomatic process more than directly alter the course of battle.[9]

Military leaders emerged from World War II retaining their belief that the utility of PSYOP resided in either persuading enemy troops to surrender or fetching deserters. Donovan's exploits with the OSS had won some converts, most notably Eisenhower, to the possibilities of PSYOP. Still, there remained too few of these individuals to move the consideration of PSYOP into mainstream military planning. In the American tradition, soldiers considered neither themselves nor their opponents as political animals. They saw little sense in expanding precious resources to develop and deliver appeals that offered them little by way of measurable results in the here and now.

The American military hierarchy, preoccupied with hardware and conventional operations during both world wars, showed little interest in integrating PSYOP into their plans. Career soldiers distrusted the civilian officials who headed the operations because they received their rank based on their communications, not their military, expertise. Service personnel perceived the whole enterprise as both temporary and outside of the military's historical concern with destroying the ability of its foes to make war. Tradition-minded officers perceived PSYOP as a field unlikely to further their professional ambitions; they strove to avoid it. With few exceptions, military use of PSYOP during the war resulted from battlefield improvisations devised in response to particular scenarios soldiers faced.[10]

American soldiers believed psychological pleas possessed usefulness only when they called the enemy's attention to the realty of conditions in the heat of battle. They preferred messages that could be used to break the fighting spirit of individuals on the other side at critical moments during a skirmish. They crafted simple and direct messages that they hoped would provide them with a tactical edge and enable them to prevail on the battlefield. They advocated statements that addressed such practical matters as prisoner treatment and wartime circumstances, not idealistic political platitudes better left to the bargaining table or memos chronicling a people's societal evolutionary status. They perceived appeals they could relate to, ones that capitalized on a soldier's individual weariness of war and rooted in a combatant's sense of truth, as the best means to entice harassed enemy soldiers either to surrender their arms or to crumble at a crucial time.

America's armed forces also adhered to the notion that psychology, at best, represented a tangential aspect of military operations that should more appropriately focus on physically annihilating the enemy. As one report observes about the traditional military perspective regarding PSYOP, "The military did not take PSYOP seriously as a primary weapon system, but believed, at best, that it was an occasionally useful supplementary weapon." The same analysis, with only slight exaggeration, notes military techniques of persuasion remained limited to yelling in the enemy's direction at the appropriate time: "Come on out or we'll blow your heads off."[11] After both world wars, the military retired or transferred to other duties its PSYOP personnel.

When the United States deployed troops to Korea in 1950, its Army possessed no plans, no doctrine, and no trained soldiers to devise or implement PSYOP. It had yet even to assess the lessons it learned during the Second World War. The Army's Office of the Chief of Psychological Warfare (OCPW), established in 1951, occupied a special staff section that reflected the military view that psychological operations constituted a subsidiary type of activity and not a fundamental element of operational strategy. For the first seven months of the war in Korea, American PSYOP consisted of one lieutenant colonel acting in a liaison capacity for the Far East Command at Eighth Army Headquarters.[12]

By the end of the conflict, though, the PWS, which came to plan and supervise all PSYOP activities in Korea, grew to include members. This development proved to be "far from smooth and regular." PWS received personnel "on assignment" from other agencies such as the Air Force and the Operations Research Office of the Department of the Army. Individuals arrived "on no schedule," and they "were of necessity fitted into the organization and into the space as seemed best when they arrived." Meanwhile, "numerous changes were made in the organizations and designations of PSYOP units in the Far East Command," most of which proved to be "short lived."[13]

True to the traditional use of PSYOP by the United States Army, the American effort in Korea orchestrated campaigns using leaflets, radio broadcasts, and loud speakers for the purpose of providing tactical support to troops engaged in fighting. Soldiers distributed materials that emphasized the misery of wartime conditions and promised good treatment to induce defections, while also accusing Russian and Chinese communists of manipulating Koreans. The main theme propounded by the Psychological Warfare Branch asserted "The Communists have launched a war of aggression to enslave all Korea and make it a puppet state."[14] This single idea became the rallying cry of American PSYOP throughout the postwar era in Asia. The United States perceived people from the Philippines to Vietnam, as even less developed than the Japanese during World War II. Against the backdrop of the Cold War's ideological contest, American leaders deemed these groups as particularly prone to outside manipulation.

Consequently, American PSYOP targeted people in both northern and southern Korea, primarily through the broadcasts of the Voice of the United Nations Command (VUNC), which began broadcasting from Tokyo on June 29, 1950, and remained on the air until June 30, 1971. VUNC came into being as a result of the efforts made by United States Army Major Thomas O. Matthews, Psychological Warfare Officer, Far East Command. In less than a day, Major Matthews gathered news copy from all of the wire services, secured translators and interpreters, obtained five transmitters, and wrote the initial thirty-minute program. Over the next two decades, the VUNC radio network came to consist of studio facilities, a newsroom, and transmitter in Okinawa, as well as a studio and a transmitter at Yongsan, Seoul, and two additional transmitters, one on Kanghwa Island (45 miles northwest of Seoul) and one at Chorwon (60 miles northeast of Seoul).

The VUNC broadcast themes were "chosen with the North Korean audience in mind," had "relevance to the Republic of Korean (ROK) audience as well." For example, a program that discussed ROK "growth and progress" served "not only as a means for causing North Korean listeners to make comparisons between conditions they were experiencing with those which they heard reported" but also aimed to instill "within Republic of Korea listeners a sense of pride in their country." Moreover, warnings aimed to North Korean audiences to refrain from renewed aggression also proved useful for "reinforcing a sense of security in the Republic of Korea audience." Other themes, such as "Communism Today," sought to "point out the failings within and without the Bloc nations," while others aimed to make clear "U.S. initiatives for peace in the Far East."[15]

WAR WITHOUT FRONTIERS

America's experience in Korea combined with the election of President Dwight David Eisenhower in 1952 to generate momentum to create a permanent agency to carry out PSYOP. By the time Eisenhower took office, the VOA reached one hundred different countries and broadcast in forty-six different languages. Concurrently, the State Department created the International Information Administration (IIA) under former Washington State University president Wilson Compton. The IIA organized United States information centers around the world, exchange programs, and the distribution of materials explaining American policies to ten thousand newspapers daily.[16]

Eisenhower entered the White House convinced of the nation's need to create a unified and strong PSYOP effort based on his observations of the OWI and OSS during the Second World War. The president's secretary of state John Foster, however, Dulles "was fairly clear that he did not wish to preside over information activities."[17] Eisenhower appointed a special committee under former CIA deputy director William Jackson, the Committee on International Information Activities (Jackson Committee), that included veterans of PSYOP during the Second World War, such as C. D. Jackson and Abbott Washburn, who served as the committee's executive secretary. The committee's central conclusion, that PSYOP needed to be integrated into America's foreign policy on an equal level with its political, economic, and military initiatives, led to Eisenhower's creation of the USIA on August 3, 1953.[18]

Officials had provided the necessary framework for integrating both information and propaganda in the USIA at the onset of its existence. On one hand, the USIA charter promulgated by the Eisenhower administration committed the agency to fostering international "harmony and understanding" by making America's case abroad. On the other hand, officials also empowered the USIA to "unmask and counter" all "hostile attempts to distort or frustrate the objectives and policies of the United States."[19] The decision to create another information agency, as opposed to a psychological warfare bureau, reflected the continued awareness of officials that they needed to maintain a rigid distinction, in the American public's mind at least between the material that the USIA disseminated and the propaganda that they accused the Soviets of peddling. So, too, did Eisenhower's choice to direct the new agency reflect this awareness. He appointed to the director's post Theodore C.

Streibert, the former chairperson of the board of the Mutual Broadcasting System radio network and president of WOR radio in New York.

Streibert publicly sang the tune carried by the former directors of America's information agencies. He shunned the notion that the USIA should ever adopt the "purely manipulative approach of totalitarian propaganda." Streibert stressed that his agency "must be content with the facts, and the facts will in the long run be the most effective arguments for our national interest."[20] He also acknowledged, nevertheless, that his agency's mandate extended beyond merely reporting the American view. He observed that the USIA mission statement "does not explicitly point to our role as a weapon of political warfare." He concluded, however, that the "hostile intentions" of the Soviet Union "impose upon the operations of the U.S. Information Agency a special character and special responsibilities."[21]

Apparently, Streibert detected no contradiction between the two strikingly different roles he identified for the agency simultaneously to play, psychological warrior and information agent. Neither did others. The USIA gained the endorsement of early public relations guru and propaganda enthusiast Edward L. Bernays and his contemporaries in marketing and communications. They lent the new agency their support through their creation of the National Committee for an Adequate Overseas Information Program. It called the USIA "a powerful offensive and defensive weapon for our nation and vital to our national strength."[22]

For the first time in its history, the United States created a single PSYOP agency that pursued both information and propaganda. This amalgamation came into being in part as a response to American perceptions of the Cold War heating up in places overseas such as Korea. It also resulted, though, from a domestic climate shaped by McCarthyism. The United States had been plunged into a deep anticommunist hysteria by Republican Senator Joseph R. McCarthy of Wisconsin. From 1950 to 1954, he accused the American government, especially the State Department's IIA, as being infiltrated by communist sympathizers.[23]

Before McCarthy's scourge finally came to an end in 1954, due in part to the nationally broadcast exposé compiled by renowned journalist and future USIA director Edward R. Murrow, the Senator's allegations wreaked havoc on the IIA and the VOA. McCarthy possessed a number of informers within the IIA that dubbed themselves the "loyal American underground." They brought about a number of brow beatings and "resignations," including those of successive IIA directors Wilson Compton and Robert Johnson.[24] While McCarthy never produced any concrete evidence of subversion within the IIA, the experience undoubtedly shaped the schizophrenia evidenced by the USIA adoption of its dual role as propagandist and information agent. The agency pledged its dedication to objectivity even as it strove to prove its unwillingness to go soft on communism.

The President's Committee on International Information Activities under William H. Jackson, former deputy director of Central Intelligence, compiled its recommendations to Eisenhower about American PSYOP during the height of McCarthy's efforts to expose communist subversives within the IIA. The group issued its findings on the day Eisenhower had imposed as its deadline to report back, June 30, 1953 a day that also happened to be the one when McCarthy requested the presence of IIA director and former president of Temple University to appear before the Permanent Subcommittee on Investigations of the Committee on Government Operations of the United States Senate. McCarthy's investigation now focused on the IIA overseas library program. The Senator wanted to take Johnson to task about "the use of books by Communist writers in this program."[25] By the end of July, Johnson submitted his resignation and returned to north Philadelphia.

The inquisition that McCarthy led on the IIA, culminating with Johnson's decision to step down, provided information personnel with an important lesson about the interconnectedness of their activities overseas with domestic politics. McCarthy's allegations, no matter how flimsy, resulted in the information program's budget being reduced by Congress from ninety-six million dollars to seventy-five million dollars. Most disturbingly, when push came to shove ,Johnson received little support against McCarthy from even the information program's chief supporters within the Eisenhower administration. According to Thomas C. Sorensen, who began working for the information program in 1951, "Although Johnson had been urged to take the job and was promised full cooperation and support, he received little from the President, the Secretary of State, or the Vice President."[26]

The report issued by the Jackson Committee on the day of Johnson's summons commented on the confusion over the purpose of American information programs. It offered, "There is still no unanimity of opinion regarding the over-all mission of the United States information agencies." While "some consider the mission to be the dissemination of truth," the task force observed, "Some emphasize the importance of winning friends for the United States, and others view the information services as a weapon against communism."[27] Rather than clarifying the program's identity, though, the committee devised a "Mission of Propaganda and Information" that endorsed the prevailing uncertainty by conflating the two activities in the mission statement's title.

The Jackson Committee called on the information program to "perform the function of informing foreign peoples of the nature of American objectives" even as it worked to "arouse in them an understanding and a sympathy" for the "kind of world order" the United States coveted. The group stressed that "objectivity should be paramount," but it also emphasized that "output should be forceful and direct both in tone and content." The committee recognized the blurriness of the line it advised that the United States information program follow. "In particular situations, propaganda, overt or covert, may play a role of decisive importance in the attainment of specific goals," it suggested, "but too much or too blatant propaganda can be harmful."[28]

The Jackson Committee's inability to establish a clear purpose for America's information program reflects the difficulties the United States has historically experienced in attempting to reconcile its need for propaganda with its democratic ideals that ward against such activities. It also should be viewed, however, in the context of McCarthyism as Americans wrestled with the question of how to best the Soviets at their own game. This point is best exemplified when one considers how the committee answered the question of what books belonged on the shelves of United States government overseas libraries. On one hand, the Committee placed its faith in American ideals by contending the program "should not hesitate to distribute books and publications just because they contain criticism" of America. On the other hand, the Committee expressed a reluctance to simply let the facts speak for themselves by adding that the overseas branches "certainly must not aid in the distribution of subversive books or communist propaganda."[29]

Historian Walter L. Hixson persuasively argues that, in the heightened anticommunist environment of the early Cold War, the United States adopted a hardcore strategy of psychological warfare as it aimed to liberate people residing under communist rule in Eastern Europe. Wherever and whenever possible, Hixson asserts, the overriding goal of the USIA involved helping to instigate rebellion against communist rule. He then suggests that, over the course of the 1950s, particularly after the failed Hungarian uprising against Soviet rule in 1956, the Eisenhower administration adjusted its PSYOP strategy by prompting the USIA to adopt a "gradualist approach." Hixson writes that decisive intervention by the Soviets in Hungary "precluded the spread of revolt to other regimes, thereby delivering a mortal blow to the elusive quest for liberation."[30] As a result, American officials overcame their "preoccupation with militarization" and utilized "a more hopeful, albeit longer term, strategy" to challenge Soviet hegemony. Through the use of straight news and the export of American culture, Hixson contends the USIA opened the door for America's gradual, but ultimately successful, "cultural infiltration of the Soviet empire."[31]

Hixson's observations are astute ones. There is no question that the USIA mellowed its approach toward Europe in the 1950s. Nevertheless, it is a simplification to suggest the transition of the USIA from a liberation strategy to a gradualist approach occurred in a complete and irreversible manner or that it applied to all areas of the world equally. The Cold War did more than prompt an initial militarization of America's PSYOP program; the conflict enabled Americans to blur the distinction that they had formally maintained between information and psychological warfare. The USIA seamlessly fluctuated between these two roles, often pursuing both simultaneously.

The subtle approach to Europe that USIA settled into by the end of the decade included the sponsorship of trade fairs and cultural exchange programs and the distribution of materials it deemed as fair and balanced. In 1956, Secretary of State John Foster Dulles viewed the program as possessing the potential to "undermine" the Soviet regime by "exposing the people to Free World influences." As early as 1949, Dulles indicated what would become the "oft-expressed view" that he shared with Eisenhower. It held that America "can only win this 'cold war' if we take the *offensive*; and that we can only win *peacefully* if we take the offensive with moral, and not merely material, weapons."[32] The USIA, infused by the spirit of Dulles and his diplomats, dramatically increased its activities. It moved to "expand, discreetly and gradually," its distribution of publications, documentary motion pictures, television and radio material, and books.[33] By the end of the decade, USIA officials boasted that their programming reflected "variety in the presentation of American opinion." They claimed it as less about "selling of soap" and more focused on providing "more honest information" to its target audience.[34]

The agency's new direction took shape under George V. Allen, whom Eisenhower named to head the agency in 1958. Allen focused his attention on improving the cultural exchange program and on enhancing the credibility of VOA broadcasts in Europe. He came to the task as a career diplomat who had held numerous ambassadorships during his twenty-nine-year career in the Foreign Service and who had also served as assistant secretary of state for information under President Truman. Allen wanted the USIA operation in Europe to mimic that of an independent news agency. He questioned whether it "is our purpose to present the government's foreign policy persuasively, or just present it and let people make up their minds about it?"[35] Former USIA deputy director Thomas C. Sorensen writes that, while Allen "was America's chief propagandist, he did not believe in propaganda."[36] Under Allen's direction, the VOA received its first formal charter, which committed it to "establish itself as a consistently reliable and authoritative source of news." It called for VOA content to remain unwaveringly "accurate, objective, and comprehensive."[37]

Allen imbued the USIA with a long-term outlook on the Cold War that exemplified the new American approach to the Cold War in Europe. He believed the agency would find the most success by exhibiting patience. Allen informed the Nevada State Press Association in 1959:

> No one of the U.S. Information Agency's activities, or all of them together, can be expected to bring about a sudden and dramatic change in our international relations. Eventually, however, a consistent, conscientious information and cultural program—fair and accurate reporting of the American scene—will surely bring about greater understanding in the world community. This, in turn, can contribute to better political and diplomatic relations.[38]

While the USIA under Allen's direction began to mold itself into a credible source of information, it also endeavored to impress on its European audience "evidence of American cultural maturity." It demonstrated a particular preference for opera and featured the form prominently in its European packets. It also emphasized the types of American "theatrical, musical and ballet events" that reinforced the notion that democracies produced a sophisticated and appealing cultural product. Henry Cabot Lodge Jr., America's representative at the United Nations under the Eisenhower administration, voiced his support for the European program to Secretary of State John Foster Dulles for a "fairly highbrow program." He specified that USIA should concentrate on exporting only "quality" music and called on the agency to "avoid jazz." Lodge stressed, "There must be no vulgarity—no matter how funny or clever or interesting the program which contains such vulgarity may be."[39]

The USIA program aimed at Europeans became less polemical by the end of the 1950s. Still, even at the beginning of the decade, the agency aimed its appeals toward the intellects and not the hearts of its audience. The "Doctrinal Program" that the agency began in 1954 strove to "break down the worldwide doctrinaire thought patterns" that provided the intellectual foundation for communism. Officials hoped for their program to "create confusion, doubt and loss of confidence in the accepted thought patterns of convinced communists." They specifically sought to influence the thinking of "intellectuals, faculties and students, historians, anthropologists, biologists, and persons in the higher circles of journalism." The agency utilized material not traceable to the United States that relied on rational analysis to "shake loose and dislodge from fixed positions the thinking of persons who have become entangled in the sophistries of Communist doctrine." Through the program, the USIA distributed books such as *What Do You Know About Democracy?* and *What Do You Know About Communism?*[40] In newly liberated former colonies around the world, though, the USIA carried out more aggressive campaigns to combat communism that assumed a high level of indigenous susceptibility to the manipulation of outsiders.

PUPPETEERS AND PAWNS

United States Army Lieutenant General William Pelham Yarborough served as Assistant Deputy Chief of Staff for Military Operations (Special Operations) from 1965 to 66. Comments he later made reflected the dominant American view that "the humble and inarticulate minority peoples of Indochina were, in actuality, very important pawns in the strategic game being played by giants in and around their homelands."[41] American PSYOP presumed that Europeans possessed the ability to rationally assess communist ideology and believed a gradualist approach would suffice. In Asia, officials could see only potential puppets that required more forceful motivation than gradualism permitted. This view is the one clearly carried by the former OSS agents dispatched by the United States to prevent the spread of communism in Asia during the 1950s and who received support from the newly created USIA.

One of these individuals, former OSS director William J. Donovan, went to Thailand as a United States ambassador. Thailand's military government, a recipient of United States aid, proved itself America's most loyal ally in Southeast Asia in the years that followed World War II. It initiated its own anti-communist publicity campaign, sent troops to Korea, and helped the CIA establish a cover for its covert plan to arm Chinese Nationalists in Burma for an invasion of southern China.[42] By the beginning of 1953, American officials sought to capitalize on Thailand's allegiance by making it a base for anti-communist operations in Southeast Asia. As one USIA official later attested, "The idea was to bolster Southeast Asia against further communist penetration, and the key was Thailand." Thai officials went along with America's planning, and the United States dispatched Donovan to Bangkok.[43]

Donovan assumed his position in Thailand and quickly laid the groundwork for USIA operations. He possessed an aggressive attitude, honed during his days as OSS chief, when he urged that for its success the United States must "steal the ball and kill the umpire" to level the playing field with totalitarian regimes. Donovan viewed PSYOP as an "instrument of war." He had been a staunch supporter of Project Fantasia, the OSS plan to inundate the Japanese in World War II with glow-in-the-dark foxes.[44] Donovan convinced Thai officials to form a joint psychological warfare committee with the United States, and he received a broad mandate from President Eisenhower to intensify and coordinate propaganda activities. Eisenhower sent him "a team of people from all over the world to help him shore up Thailand against communist subversion." Under Donovan's supervision, the USIA opened libraries, broadcast radio programs, held exhibits, and presented movies.[45]

The USIA initiated its program for the "anti-communist indoctrination of the whole national structure" by presenting an exhibition in Bangkok that exposed "Communism's record of crimes in Asia and Europe." Americans had set out to ensure that the Thais, a people they cast with other Asians to "place great faith in supernatural beings," received the insulation against communist duplicity they required.[46] The agency estimated that 180,000 people attended the exhibit, officially sponsored by Thailand's national police. In May 1954, USIA began its "intensive indoctrination of the government apparatus" by offering a formal six-week course for 28 senior Thai officials. Thailand's police general Phao Sriyanond addressed the officials and stressed the program's significance for his nation's internal security. He also emphasized the "necessity of continuing cooperation and understanding" between Americans and Thais "in the face of the Communist danger." USIA officials considered their foray to protect Thais from outside manipulation, "greatly assisted by the U.S. Ambassador," a successful venture.[47]

The United States also acted quickly in other parts of Asia, where it detected communist uprisings already underway. The Hukbalahap rebellion, an armed guerilla revolt in the Philippines that erupted in 1946 and crested around 1950, attracted the attention of American policymakers.[48] The Huk uprising originated with the Filipino peasantry in the Luzon plain, who suffered under oppressive landowners who controlled ninety-eight percent of the land. Landlords demanded rents from tenants as high as fifty percent of the land's total yield in rice. Initially aimed at the Japanese during their occupation of the Philippines during the Second World War, the Huk movement continued its resistance against the postwar Filipino government of President Manuel Roxas, who had collaborated with the Japanese.

By the summer of 1950, against the backdrop of war in Korea, the United States rushed to aid the pro-American government in Manila for fear that otherwise a pro-communist regime could take power. As American money poured into the Philippines to enable Manila to check the allegedly communist-led Huk threat, the Joint U.S. Military Assistance Group (JUSMAG) deemed the Filipino unrest to constitute an unconventional conflict. It called on Washington to dispatch special advisory personnel to address the situation. Former OSS operative Edward Geary Lansdale, who now served as a Lieutenant Colonel in the United States Air Force, arrived in the Philippines in September 1950 in order to take charge of the effort against the Huk.

At the same time that Lansdale arrived in the Philippines, Ramon Magsaysay, a former Filipino anti-Japanese guerilla, became the new Minister of National Defense in charge of counterinsurgency. Besides his efforts against the Japanese, Magsaysay possessed an outstanding reputation among American officials. The United States Army had appointed him as a provincial military governor following Japan's ouster. Moreover, Magsaysay had visited the United States several times in the 1940s to secure veteran's benefits for Filipinos and to petition for greater aid against the Huks. Lansdale quickly became Magsaysay's confidant and advisor, ultimately helping his election to the Filipino presidency in 1953.[49]

Possessing nearly unlimited authority in the Philippines, Lansdale set up the Civil Affairs Office (CAO), a PSYOP division he and Magsaysay controlled. The CAO engaged in any number of anticommunist activities, from funding student organizations to coordinating with the USIA in the production and distribution of over ten million leaflets. Most important, though, the campaign contributed vitally to the Huk rebellion's collapse by 1954. Lansdale and Magsaysay utilized the agency to mount an enormous public relations campaign on behalf of the government that successfully presented the appearance of government reform.

According to Armed Forces of the Philippines retired Colonel Napoleon D. Valeriano and United States Army retired Colonel Charles T. R. Bohannan, two of Lansdale and Magsaysay's officers in the Philippines, "A major mission of the CAO was to secure the widest publicity for all laudable accomplishments of the armed forces." The organization constantly "fed" information "to eager newspaper reporters and radio commentators." It publicized, for example, the dispensation of candy to children by Filipino forces as it moved into areas to clear Huk, a crucial part of Magsaysay and Lansdale's "Attraction Program."[50]

The CAO achieved its greatest publicity strike against the Huk in guiding and promoting the Economic Development Corps (EDCOR), Magsaysay's effort to deliver on his promise of providing "Land to the Landless." Captured Huk "who gave evidence of their sincerity" received about fifteen acres of land to work that could then be purchased outright when the harvest made payment possible. Valeriano and Bohannan judge the program "a tremendous success" as a psychological operation and note Magsaysay estimated that "the value of this operation" stood as "greater than that of another battalion combat team in the field." Still, it is clear the program amounted to more in appearance than it did as real reform. In total, only two hundred and forty-six Huk actually benefited from the program. Even Valeriano and Bohannan concede that EDCOR projects "did not in themselves make a significant contribution to the economic or social welfare of the country."[51]

To officials in Washington, Lansdale's efforts on behalf of Magsaysay appeared easy. They stood mesmerized by Lansdale's successful use of an unconventional "smoke and mirrors" approach through programs such as EDCOR and the "psy war tactics" of his preference, "practical jokes." Lansdale honed his technique on a Filipino culture he defined as exotic and "superstitious." He shows an endless interest in traditional customs, noting in his memoirs, "When I introduced the practical-joke aspect of psywar to the

Philippine Army, it stimulated some imaginative operations that were remarkably effective." The most creative scheme capitalized on a Filipino belief in vampires in order to flush out a previously unflappable Huk squadron entrenched in a hill town. To initiate his plan, Lansdale planted stories of a vampire sighting among the local population. He then acted to provide the rumor with teeth. His Filipino charges captured and killed a Huk, "punctured his neck with two holes, vampire-fashion, held the body up by the heels, drained it of blood, and put the corpse back on the trail." Of his exercise in the macabre, Lansdale boasts, "When daylight came, the whole Huk squadron moved out of the vicinity."[52]

Lansdale's efforts on behalf of Magsaysay's government perpetuated the American notion of Asian susceptibility to superstition and foreign manipulation. Lansdale seemingly made use of every outlandish idea that occurred to him, from the display of mutilated bodies to the use of "magic eye" informants and loudspeaker broadcasting of "the Voice of God." Most of these "tactics" would be replicated in Vietnam.[53] While Lansdale publicly credited Magsaysay's leadership in bringing the Huk rebellion to a close, he made no secret off the record that he considered the Filipino leader as both malleable and superstitious.[54]

A VIETNAMESE MAGSAYSAY?

At the same time Huk rebellion collapsed under the Magsaysay initiative in the Philippines, American officials turned their attention to Southeast Asia, where France, a key American Cold War ally in Western Europe, struggled to hang onto its colonial hold over Vietnam. In January 1954, the Presidential Special Committee on Indochina considered the options for American intervention. Among other proposals, it suggested providing the "anticommunist" Vietnamese with "an unconventional warfare capability."[55] Fresh from his successful adventure in the Philippines, now Colonel Edward Geary Lansdale headed to Vietnam to take charge of the Saigon Military Mission (SMM).

Lansdale's effort did not constitute the first instance of American PSYOP in Vietnam. His former agency, the OSS, operated there during the Second World War against the Japanese. During that time, American actions supported the very group that the French and their Vietnamese allies now opposed. In 1930, Ho Chi Minh, born Nguyen Sinh Cuong in 1890, helped found the Indochinese Communist Party (ICP) to lead a revolution against French colonizers who had occupied the country since the middle of the nineteenth century.[56] Ho, by this point, had acquired an international reputation as a Vietnamese nationalist. He left Vietnam to work on a French ship in 1911, settled in Paris, and became politically active on behalf of his colonized nation. He demanded autonomy and democratic rights for the Vietnamese at the Paris peace talks that followed the First World War (1914–1918).

After being ignored by the Allied leaders at the Versailles conference, Ho became a founding member of the French Communist Party (FCP). He likewise became an agent for the Comintern, the Soviet apparatus for provoking an international revolutionary movement. He studied in Moscow in the early 1920s before heading first to South China then back to Vietnam. The ICP Ho helped to create there aimed to establish an independent socialist state. While Ho's organization did not achieve its goal immediately, it managed to survive and to seize the opportunity provided by World War II. Following Nazi Germany's defeat of France in 1940, the collaborationist Vichy government cooperated with Japan, allowing it to exploit Vietnamese resources. In 1941, ICP leaders created the Vietnam Independence League, or Viet Minh, comprised of both communist and other nationalists. The group worked with the Allies as its members strove to rid their country of a foreign presence once and for all.

By the time the Viet Minh came into being, the old ICP had been decimated in southern Vietnam by French suppression after the party headed an abortive insurrection in December 1940. Much of its leadership in that part of the country was jailed or executed, a circumstance with great consequences for the newly created Viet Minh. It facilitated the ascendancy within the new party of leaders such as Ho Chi Minh, who, by this point, had come to stress the nationalist component of the ICP at the expense of the call for immediate socioeconomic revolution. The creation of a socialist state now firmly took a back seat to facilitating the formation of an independent Vietnam.[57]

Following Japan's defeat in 1945, Ho acted to establish the Democratic Republic of Vietnam. He appealed to the international community to recognize the new regime. In particular, Ho hoped to receive America's endorsement, and, from his perspective, every indication pointed in the direction that he would receive it. In addition to the OSS cooperation, he received during the war, Ho received a show of support from his American colleagues once the fighting had ended. Some of them attended the ceremony Ho held to declare Vietnam's independence on September 2, 1945. In addition, in conjunction with the OSS, United States Army General Philip Gallagher, advisor to the incoming Nationalist Chinese occupying force, created the Vietnamese-American Friendship Association. Perhaps most important, though, as pointed out by the late historian George McT. Kahin in his study on America's involvement in Vietnam, *Intervention*:

> Like those of anticolonial nationalists in other Southeast Asian countries, his [Ho's] expectations had been affected by several other factors: the persistent advocacy of the principle of self-determination in the wartime broadcasts of the

> U.S. Office of War Information; the United States' unqualified espousal of the Atlantic charter's provision that sovereign rights and self-government would be restored to all peoples who had been forcibly deprived of them; and Washington's clear-cut promise of independence to the Philippines.[58]

France, however, sought to return to its imperial position in Vietnam. It convinced the American government of the necessity for its postwar economic stability of continuing its prewar status in Southeast Asia. To bolster its case and promote the appearance of its commitment to initiating some political reforms, France extended nominal independence to Vietnam by installing Nguyen Vinh Thuy, the Bao-dai Emperor who ruled from 1926 to 1945 and who cooperated with Japan during the war, as the chief of a State of Vietnam (SOV). The United States, under both Truman and Eisenhower, supported the French and their creation. In part, they were motivated by the need for postwar stability, but no less influential was the domestic political pressure that resulted from the fall of China in 1949. No American leader could now appear soft on communism.[59]

By the time Lansdale arrived, American policy had focused on creating an alternative to Ho's Democratic Republic of Vietnam (DRV). American officials saw it as in their nation's best interests to make more lasting the temporary line of demarcation along the 17th latitudinal parallel brought Vietnam by the DRV-French truce established by the Geneva Accords. American officials feared that, if the reunification elections that the Geneva Accords called for took place as scheduled in 1956, Ho would win the day based on his status as a national hero. Vietnam would become part of a growing Asian communist bloc extending south from China.[60] With the USIA playing a key advisory role, the United States quickly acted to provide the Vietnamese with an American-friendly government they hoped would allow them the time to find a Vietnamese nationalist hero of their own. They turned to Bao Dai and the remnants of the SOV leadership to establish an anticommunist state based in Saigon and in control of territory south of the 17th parallel.

Officials believed Ngo Dinh Diem, anti-French and anticommunist, represented America's best answer to the communist challenge they detected in Vietnam. They lent him their backing, despite at least some skepticism existent in the Eisenhower administration regarding his commitment to bring about socioeconomic reforms.[61] In 1955, Lansdale hit the ground in Vietnam running. He precipitated a massive movement of people into the southern part of the country by circulating leaflets that warned of a draconian new social order that purported to be signed by officials in Hanoi. The USIA produced stories and photographs of this exodus of nearly one million people to illustrate it as a tribute to democracy and an indictment of communism, even as

Lansdale worked to convince undecided northern Catholics that Jesus Christ had gone south. They needed to follow him there.[62]

As the refugees streamed in from the north, Lansdale set to work in Saigon. He pulled from his old Filippino bag of tricks and hired astrologers who foretold disaster for "certain Vietminh leaders and undertakings." Lansdale circulated rumors of a marauding Chinese force in the north and disseminated pamphlets of a "Thomas Paine type series of essays" extolling the patriotic virtues of resisting communism.[63] By March 1955, the USIA Motion Picture Production Section in Saigon began to produce, on average, one documentary per month on behalf of Diem's leadership. Over the course of the next three years, the section increased that output to three films each week. It distributed movies through the Vietnamese Ministry of Information, which screened them in mobile film units throughout the countryside. A typical film, *Bird in a Cage*, according to former Motion Picture Officer Myriam L. Johnston-Hallock, constituted a "sugar coated anti-communist pill arranged in a poignant tale."[64]

Under the instructions to do what he did in the Philippines, Lansdale, aided by the USIA and the CIA, went about creating a southern Vietnamese state under Diem's rule. The American team provided Diem with sufficient financial backing to eliminate, or to bribe, organized opposition to Diem in 1955. Diem initially had alienated the leadership of the religious Cao Dai and Hoa Hao sects by refusing to recognize the autonomy of the territories that these groups administered. He proceeded to cut their funding and limit their representation in Saigon. The promise of American financing brought these groups into the fold, which enabled Diem to squash the Binh Xuyen, a group of brigands that under the French had gained control over much of Saigon's police force and operated unchecked a lucrative ring of organized crime.[65]

Following Diem's consolidation of power in 1955–56, Lansdale and his team did all in its power to create the impression of the new government's legitimacy. The American effort resulted in first a referendum then an election, as a means to project Diem's RVN as a representation of Vietnamese democracy. Of course, the Lansdale group did all it could short of being accused of outright fraud to ensure the results of each vote firmly secured Diem's leadership. The USIA, through the posts it opened in Saigon, Can Tho, and Dalat, and through the advice it provided to the new Vietnamese Information Service (VIS), carried out an enormous public relations campaign on Diem's behalf.[66] During the actual voting, the American team carefully crafted the language and even the color of the ballots used to bring about the desired outcome of a Diem presidency.[67]

In power and in possession of American backing, Diem went along with the idea of democratic government in appearance, but he kept real authority in his own firm possession. Through a covert political apparatus, the Can Lao party, one that Diem ran through trusted family members such as his brother Ngo Dinh Thuc, the Catholic archbishop of Hue and primate of Vietnam, the Diem regime stifled all opposition to its rule. The Diem government banned opposition newspapers and "wreaked its wrath against the press" through "well-organized mob action." It dissuaded dissenters through means such as "surveillance of their movements" and "sacking of their places of business." The regime further resorted to arresting its opponents "on trumped-up charges."[68]

Diem justified much of this oppression as a necessary measure against communism. In the end, however, the creation of a "quasi-police" state reminded its primary target, the Vietnamese peasantry, more of French repression than the promise of a free Vietnam.[69] Diem abolished elections for village councils, supported landlords in helping them repossess land that had earlier been seized from them by the Vietminh and turned over to peasants, increased land rents, and began a process meant forcibly to assimilate ethnic minorities such as the Montagnards, who resided in the interior highlands north of Saigon. Only in 1958, after "continued prodding by U.S. officials," did Diem finally introduce a program of land distribution. This initiative, though, "was so emasculated and rigged in favor of landlords that it usually inflamed rather than assuaged peasant hostility." By the end of 1962, only about ten percent of the one million tenant households in southern Vietnam had obtained land titles.[70]

American Vietnam War correspondent Neil Sheehan observes "South Vietnam, it can truly be said, was the creation of Edward Lansdale."[71] In the sense that Lansdale taught Diem all of the "tricks" he employed in the Philippines and encouraged his preferred "Vietnamese Magsaysay" to act on his own oppressive impulses to thwart communism, Sheehan's observation rings true. Lansdale's "creation" brought the appearance of stability in southern Vietnam. Still, it planted the seeds for profound discontent in the southern countryside. In her synthesis of the Vietnam War, historian Marilyn B. Young persuasively argues that Lansdale's "cheerful brutalization of democratic values" did far more harm than good to the RVN cause.[72]

Lansdale's rumor campaigns and practical jokes rarely supplanted the allegiance that Ho Chi Minh had accumulated with his nationalist message. Instead, Lansdale's initiatives often reflected back on the RVN and illuminated the discrepancy that existed between its democratic rhetoric and its undemocratic means of maintaining power. In his dual obsession with usurping Ho Chi Minh's standing and stabilizing Diem's position by all means possible, Lansdale failed to consider the RVN's role in undermining its own standing in the hearts and minds of the Vietnamese who linked such oppressive tendencies with collaborationist regimes.

Lansdale's practical jokes, no matter how cleverly conceived or executed, could not prompt those who sympathized with the Vietminh before his arrival to back a government in Saigon that remained woefully out of sync with its intended constituency. Already the Vieminh had effectively appropriated Vietnamese history to its advantage. It had deeply rooted itself within Vietnamese imaginations by evicting the French colonial regime.[73] Unlike Lansdale, who believed it possible to create nationalism in Asia from scratch by employing the correct combination of rhetoric, trickery, and stabilization, the leadership of the Vietminh had appreciated it as a historical force that required cultivation through both deeds and action.

From the perspective of Vietnamese communist party leadership, by 1956, "events were not going well."[74] Diem, with American support, had managed to scuttle the 1956 national election mandated by the previous year's signing of the Geneva accords. In addition, as opposed to collapsing as Lao Dong leaders had initially expected, the Diem regime apparently had succeeded in consolidating its power by eliminating the sects, the emperor Bao Dai, and a growing amount of the opposition that remained. Party members in the nation's south grew impatient with the inaction that characterized their organization's stance against Diem.

The reasons behind the Party's caution in endorsing open revolt in the south involved the need to strengthen the nation's north. The Lao Dong leadership "was acutely concerned" that widespread armed resistance in the south might "precipitate direct U.S. military intervention" on Diem's behalf.[75] Consequently, the Party responded in 1956 by urging patience, but it also called for political activity, which included not only propaganda appeals to Vietnamese nationalism, but also endorsement of the ultra secret "extermination of traitors movement" aimed at eliminating government "hard-core" elements.[76]

Political agitation and assassination are a volatile mix, however. Over the course of the next three years, the Lao Dong leadership found itself attempting to restrain, or at least provide some guidance to, the increasingly restless masses in the south that sought to force the issue with Diem and his American supporters. By 1960, it became clear to Party officials that their cautionary approach needed to be abandoned as the threat surfaced that a full-scale assault against Diem could potentially erupt without their oversight. In March of that year, a group of "former resistance fighters" called for the commencement of a "struggle to put an end" to the Diem government and to rid the nation of American influence. It advocated the application of the Geneva

Accords, the creation of a coalition government to replace Diem, and the peaceful reunification of Vietnam.[77]

Initially, the Lao Dong leaders expressed reluctance to endorse this new bold plan of action because of their continued wariness of American intervention. It had become evident, though, that by this point they possessed little choice. The southern discontent with inaction did not dissipate. In September 1960, the Third Lao Dong Party Congress officially encouraged southerners to take direct action against Diem by calling on them to create a broad National United Front that could oust the government in Saigon and its American backers.

By the end of the year, southerners met to create a formal organization committed to this task. On December 20, the National Front for the Liberation of South Vietnam (National Liberation Front, or NLF) came into being as a result of this southern initiative. The organization proposed a ten-point program that assigned top priority to the removal of the Diem government, which it identified as a product of American colonialism. With the Party's public announcement of the founding of the NLF in January 1961, and the creation a year later of a People's Revolutionary Party (PRP) to instill the Front with an avowedly communist element, the southern revolution entered into a "new and more bitter stage."[78]

NOTES

1. Stephen E. Ambrose and Richard H. Immerman, *Ike's Spies: Eisenhower and the Espionage Establishment* (Jackson: University Press of Mississippi, 1981), 163.
2. Ambrose and Immerman, *Ike's Spies*, 165.
3. Hixson, *Parting the Curtain*, 17.
4. See Rhodri Jefferys-Jones, *The CIA and American Democracy* (New Haven: Yale University Press, 1989); Mickelson, *America's Other Voice*; Nelson, *War of the Black Heavens*; and Puddington, *Broadcasting Freedom.*
5. Henderson, *U.S. Information Agency*.
6. Hixson, *Parting the Curtain*, 11.
7. Barrett, *Truth is Our Weapon*; Hixson, *Parting the Curtain*, 15.
8. Laurie, "'The Chanting of Crusaders.'"
9. Laurie, *The Propaganda Warriors*; Bradley F. Smith, *The Shadow Warriors*; Harris R. Smith, *OSS: The Secret History*.
10. Daugherty and Janowitz, *A Psychological Warfare Casebook*.
11. Report, Psychological Operations, undated, "Conduct of the War: Functional Analyses Volume VI, Book 2 [3 of 3]," box 4, BDM Corporation Study: A Study of Strategic Lessons, LBJL.

12. William E. Daugherty, "PSYOP in Perspective," in *The Art and Science of Psychological Operations*, I: 80–85.
13. PSYOP—POLWAR Newsletter, Voice of the United Nations Command: A Description of a Strategic Radio Broadcasting Operation, August 1, 1971, folder 8, box 14, Douglas Pike Collection: Unit 03—Insurgency Warfare, VA-TTU.
14. Report, "Psychological Warfare in Korea," *Public Opinion Quarterly* 15, no. 1 (spring 1951): 65–75.
15. PSYOP—POLWAR Newsletter, August 1, 1971.
16. Hixson, *Parting the Curtain*, 21.
17. Oral history, Theodore Streibert, in *The United States Information Agency: A Commemoration* (Washington, D.C.: Government Printing Office, 1999), 20.
18. The Report of the President's Committee on International Information Activities, June 30, 1953, *FRUS, 1952–1954*, II: 1836.
19. Mission of the United States Information Agency, undated, *FRUS, 1952–1954*, II: 1753.
20. U.S. Information Agency: Principles, undated, *FRUS, 1952–1954*, II: 1763.
21. U.S. Information Agency: Principles.
22. *The United States Information Agency*, 15.
23. Hixson, *Parting the* Curtain, 52–54; David M. Oshinsky, *A Conspiracy So Immense: The World of Joe McCarthy* (New York: The Free Press, 1983); Sorensen, *The Word War*; 31–35.
24. Sorensen, *The Word War.*
25. The Chairman of the Permanent Subcommittee on Investigations of the Committee of Government Operations of the United States Senate to the Administrator of the International Information Administration, June 30, 1953, *FRUS, 1952–1954*, II: 1718.
26. Sorensen, *The Word War*, 36.
27. The Report of the President's Committee on International Information Activities, *FRUS, 1952–1954*, II: 1838.
28. The Report of the President's Committee on International Information Activities, 1836.
29. Sorensen, *The Word War*, 42.
30. Hixson, *Parting the Curtain*, 82.
31. Hixson, *Parting the Curtain*, *227*.
32. Bowie and Immerman, *Waging Peace*, 52.
33. See Hixon, *Parting the Curtain*, 108–10.
34. Barry Zorthian, "Remarks," in Hans N. Tuch and Lewis Schmidt, eds., *Ike and USIA: A Commemorative Symposium*, (Washington, D.C.: The United States Information Agency Alumni Association, 1991), 20.
35. Nelson, *War of the Black Heavens*, 102.
36. Sorensen, *The Word War*, 116.
37. Sorensen, *The Word War*, 235.
38. Henderson, *The United States Information Agency*, 21.
39. Memo, The United States Representative at the United Nations (Lodge) to the United States Secretary of State (Dulles), July 23, 1954, *FRUS, 1952–1954*, II: 1775.

40. United States Doctrinal Program, January 15, 1954, *Declassified Documents Reference System* (Woodbridge, Connecticut: Primary Source Media, 1993), 508, F14–19, microfilm.

41. William P. Yarborough, "Civil Assistance in Laos," in *The Art and Science of Psychological Operations*, I: 458.

42. See Daniel Fineman, *A Special Relationship: The United States and Military Government in Thailand, 1947–1958* (Honolulu: University of Hawaii Press, 1997).

43. James J. Halsema, "Remarks," in *Ike and USIA*, 32.

44. Deuel, *History of OSS*, I: 105.

45. Halsema, "Remarks," 32.

46. Imogene E. Okes, "Effective Communication by Americans with Thai," in *The Art and Science of Psychological Operations*, II: 594.

47. NSC 5430, Part 7—The USIA Program, August 12, 1954, *Declassified Documents*, F10–12.

48. Charles Bohannan and Napoleon Valeriano, *Counterguerilla Operations: The Philippine Experience* (New York: Frederick A. Praeger, 1962); Benedict J. Kerkvliet, *The Huk Rebellion: A Study of Peasant Revolt in the Philippines* (Quezon City, Philippines: New Day Publishers, 1979); Stephen R. Shalom, "Counter-Insurgency in the Philippines," in Daniel Shirmer and Stephen R. Shalom, eds., *The Philippines Reader* (Boston: South End Press, 1987).

49. Michael McClintock, *Instrument of Statecraft: U.S. Guerilla Warfare, Counterinsurgency and Counterterrorism, 1940–1990* (New York: Pantheon Books, 1992).

50. Charles Bohannan and Napoleon Valeriano, "Target—The Civilian," in *The Art and Science of Psychological Operations*, I: 443.

51. Bohannan and Valeriano, "Target—The Civilian," 445.

52. Lansdale, *In the Midst of Wars,* 72.

53. Bohannan and Valeriano, "Target—The Civilian," 440.

54. McClintock, *Instrument of Statecraft*, 95.

55. "Lansdale Team's Report on Covert Saigon Military Mission in 1954 and 1955," undated, in *The Pentagon Papers,* Gravel edition, I: 579.

56. See Kahin, *Intervention*; Kolko, *Anatomy of a War*; McLeod and Nguyen, *Culture and Customs of Vietnam*; Neil Sheehan, *A Bright Shining Lie: John Paul Vann and America in Vietnam* (New York: Vintage Books, A Division of Random House, Inc., 1989); and Young, *The Vietnam Wars*.

57. Kahin, *Intervention*, 10–14.

58. Kahin, *Intervention*, 14.

59. Kahin, *Intervention*, 31–32.

60. Kahin, *Intervention*, 89; Spector, *Advice and Support*, 220.

61. Kahin, *Intervention*, 80; Seth Jacobs, *America's Miracle Man in Vietnam: Ngo Dinh Diem, Religion, Race and U.S. Intervention in Southeast Asia, 1950–1957* (Durham, North Carolina: Duke University Press, 2004).

62. Sheehan, *A Bright and Shining Lie*, 136.

63. "Lansdale Team's Report," *The Pentagon Papers,* Gravel edition, II: 573–83.

64. Henderson, *The United States Information Agency*, 19.

65. Kahin, *Intervention*, 83.

66. Report, Administration History: USIA Volume I (2 volumes), 1968, "Volume I: Administration History [1 of 2]," box 1, Administration History, LBJL.

67. Bernard Fall, *The Two Vietnams: A Political and Military Analysis* (New York: Praeger, 1964), 257; Kahin, *Intervention*, 96.

68. Scigliano, *South Vietnam*, 89.

69. William Henderson, "South Vietnam Finds Itself," *Foreign Affairs* 35, no. 2 (January 1957): 285.

70. Kahin, *Intervention*, 99.

71. Sheehan, *A Bright and Shining Lie*, 138.

72. Young, *The Vietnam Wars*, 46.

73. Race, *War Comes to Long An*; Sheehan, *A Bright and Shining Lie;* Young, *The Vietnam Wars*.

74. Race, *War Comes to Long An*, 74.

73. Kahin, *Intervention*, 105.

75. Race, *War Comes to Long An*, 83.

76. Kahin, *Intervention*, 114.

77. Kahin, *Intervention*, 114.

78. Race, *War Comes to Long An*, 123.

Chapter Six

A Matter of Performance

OVERVIEW

This chapter consists of five sections. The first, "Mind the Gap," indicates that by the end of the 1950s Americans had grown alarmed over the gap they detected between their own PSYOP and those carried out by their communist adversaries in the developing world. They considered people there to lack a sufficiently strong sense of nationalist identity to resist anti-Western propaganda. The second section, "Organize and Mobilize," illustrates that as Americans considered the role of their PSYOP program in their "nation building" efforts, they refused to recognize the subjectivity of their nation's global perspective. Consequently, to improve on their PSYOP program, Americans focused on sharpening its organization and technique, not its message.

"A New Frontier," the chapter's third section, chronicles the efforts made by the John Fitzgerald Kennedy presidential administration to retool the USIA. Under the direction of Edward R. Murrow, the renowned journalist, America's PSYOP maintained the outward of appearance of "objectivity." However it began to pursue a more aggressive initiative to increase American prestige overseas. The fourth section, "A Sense of Urgency," demonstrates that anticommunism and racism drove American assessments of Vietnam. Despite evidence to the contrary, officials considered unrest there as a product of external agitation that southern Vietnamese lacked the will to resist without American help.

This chapter's last section, "Saving Face," focuses on the crisis that Americans faced in Vietnam in 1963 when Buddhist protests against the oppressive policies of the Diem administration underscored the gap between the democratic appearance and the dictatorial rule of the RVN. Embarrassed by the repression ordered against the Buddhist movement by Diem, but also alarmed by the willingness of American correspondents to cast American support for the RVN in an unfavorable light, United States officials proved increasingly willing to use the USIA in Vietnam to influence the creation of a more positive portrayal of the Saigon government in media reports.

MIND THE GAP

Even as indigenous opposition to the American-backed Diem regime intensified in the late 1950s and carried into the 1960s, the United States marginalized local participation in "communist insurgencies" directed against "democratic governments" that it supported in the "developing world." In 1958, American officials charged by Congress with the oversight of the USIA identified communist propaganda from Moscow as the "boldest, most successful challenge to the United States and the free world."[1] They worried about the Soviet Union's rededication to ideological struggle and believed its launch of the Sputnik satellite widened an already substantial communist advantage in the competition to impress impressionable hearts and minds of people in developing countries. The Soviets, they believed, had "reached and exceeded the capitalist countries in this field to which they had given priority." These officials estimated that the United States stood "30 years behind in the competition with Communist propaganda" at the Cold War's onset, and they urged still "the gap has not been closed."[2]

Two years later, officials predicted the upcoming decade would be filled by unprecedented ideological warfare throughout the world instigated by communist leaders in Moscow and Beijing. "Competition between communism and the free world," they projected, "will be increasingly intensified in the underdeveloped areas of the world." Officials advised that the greatest challenges facing the United States "may well occur in South and Southeast Asia, Africa, the Middle East, and Latin America," where they foresaw "the strongest efforts to influence minds, and swing opinions."[3] They expressed no doubt that "there is top level control and direction of the Communist propaganda apparatus," and "each year there has been continuous enlargement of its program and facilities."

American officials concluded "There is every reason to believe this expansion will continue in the decade ahead, especially as the Chinese Communist flow has just begun." They "urgently" recommended that the United States engage in a "constant effort to improve our [American] *quality* of output." America could not match the communist world in terms of its quantity of output. "To keep pace, we [the United States] must *outperform* them [communists]."[4] The PSYOP program, from the top down, subscribed to and reinforced the notion that external communist agitation constituted the primary reason for domestic unrest in the "undeveloped world." Despite USIA engagements throughout the 1950s in Asian societies such as Vietnam, the Philippines, and Laos, the program proved either unwilling or unable to detect grievances that manifested locally to bring about indigenous action.

While both the Soviets and the Chinese offered their support to indigenous "national wars of liberation" against what they labeled as American neocolonialism, historians Lawrence J. Bassett and Stephen E. Pelz suggest American officials should have been more critical in assessing the support offered to the Vietnamese by these two powers. They point out, for example, the Chinese showed no enthusiasm for reviving the Vietnamese insurgency before 1960 and had been involved in moving Ho Chi Minh to accept the division of Vietnam in the first place at Geneva in 1954. Meanwhile, Soviet Premier Nikita Khrushchev's endorsement of national liberation wars in a 1961 speech also proposed increased trade between the East and the West and the need for disarmament negotiations. According to Bassett and Pelz, "Neither the Chinese nor the Russians wanted to press national liberation wars to a direct superpower confrontation, and both recognized the centrality of indigenous sources to revolutionary success." American officials strained to detect the difference between the real and rhetorical support offered to insurgencies by the communist superpowers. They cast Vietnam into the role of a pawn, though, with considerable ease.[5]

This action stemmed from the same racialist assumptions Americans had exhibited in waging PSYOP against the Japanese during the Second World War. "What kind of information program," such Americans as retired USIA Foreign Service Officer (FSO) James D. McHale asked themselves, made sense in "such a primitive, high risk environment" as a place like Laos or Vietnam? The answer they devised, of course, involved "nation building." This process involved two primary components that grew out of the idea that these people stood vulnerable to outside manipulation because of their "developing" status. McHale, who served in Laos for six months in 1959, reveals the first step needed as "security for the province." The second element, according to fellow retired FSO Yale Richmond, who spent two years in Laos (1954–1956), involved literally giving people a national identity. He remembers, "Most people did not know they had an independent state;" consequently, the American job consisted of "nation building from the ground up."[6]

ORGANIZE AND MOBILIZE

Throughout this endeavor of nation building, American officials never doubted the impartial perspective on international affairs they believed their nation to enjoy. In March 1962, a symposium convened at Princeton University to consider the questions raised by the USACI identification of the propaganda gap between the United States and its communist adversaries. Its participants consisted of academics, government officials, and representatives from the private sector who shared a conviction that information policy played a vital role in international relations and who also believed America's program required a re-examination.

John B. Whitton, founder of the Princeton Listening Center for the Study of Political Broadcasting, opened the meeting by proclaiming it "obvious" that PSYOP constitute "a major factor in the Cold War." He continued, "But the American effort in this field, according to many competent authorities, has been far from effective." Whitton contended that America's program "has not been a success." He alleged that Americans stood "confused as to its purpose. We lack faith in its effectiveness. We have remained too much on the defensive."[7]

Other conference participants who followed Whitton differed widely about how America's program should operate to become more efficient. Some proposed that the USIA should not veer from its established detached and empirical presentation. George V. Allen, USIA director (1957–1960) contended, all Americans can do "is to put forward as honest, objective, and truthful an information program as God gives us to see the truth." He believed all he and his fellow Americans could do meant to "rest our case with the common sense of mankind." Officials needed to possess "a mystic faith, as Jefferson did," Allen observed, "in the ability of the common man to make a right decision if given adequate information."[8]

For all of Allen's idealism, he did not address the question others at the conference harbored. What if, given "adequate information," the "common man" did not make the right decision as American policy makers perceived it? Robert T. Holt, a political science professor from the University of Minnesota, suggested that the United States must adopt a more persuasive approach than the traditional one. He proposed American PSYOP should aim to "manipulate people's perception of the world independently or relatively independently of any changes in the material world." USIA activities, he argued, "must be justified by the fact that they will contribute to the eliciting of certain kinds of acts from decision makers of foreign states-acts that will be directly related to the achievement of foreign policy goals." Holt labeled the "information approach" to PSYOP as "woefully insufficient." The USIA operating only as a "worldwide news service," he suggested, "can never be used to its full potential."[9]

Finally, former CIA director Allen W. Dulles advised that at times Americans needed to balance their own needs against the way others perceived their actions. He urged, "In the interest of our security and world peace, from time to time we have to do things that some people do not like." Dulles believed

Americans "cannot expect" that they "can make all of the people of the world love us all the time." Still, he too held firm that American objectivity would bring a vast many individuals around to the side of the United States. While conceding that Americans "have something to learn from communist propaganda techniques," he saw "no reason either to emulate them or ascribe them a power" that he thought they did not possess. He labeled communist tactics as "patent lies" that earned credibility only through endless repetition by an American press he accused of running propaganda "as though it were a fact."[10]

Despite their differences regarding USIA strategy, none at the conference questioned America's just and impartial perspective on foreign affairs. Those who criticized the USIA approach did so because they had become skeptical about the rest of the world's ability to detect America's honest intentions without receiving more forceful prodding. Robert Holt, the most outspoken critic of USIA at the meeting, noted, "American policies since the end of the Second World War have been marked by a real nobility of purpose." The United States had refused to "establish a world imperium in 1946 when we alone had mastered the terrible secret of the atom," he pointed out. When America failed, he concluded, "It has not been because our goals were improper, but because our strategies were inadequate."[11]

Officials refused to concede that it remained possible to raise legitimate questions regarding the American perspective on global events. They believed if a group of people did not concur with United States policy, only a breakdown in communications, which they ascribed to either the method of delivering the message or their audience's perfidy, could explain the difficulty. As the United States Advisory Commission had stressed in its Fifteenth Annual Report on the PSYOP program, "the essential task" for the USIA involved its ability "to marshal and organize resources more effectively." The report emphasized, "If the ideas of freedom are to prevail, they must be mobilized, organized and communicated to the peoples of the world."[12]

This notion provided Americans with an explanation for any problems they encountered in communicating United States policy persuasively. They understood the successful outcome of their efforts as dependent on the effective organization and technical delivery of America's message—not on the policy itself. The process of communication, as opposed to the actual material they constructed based on the presumed objective perspective of the United States, was the aspect they considered as paramount to their success.

At the Princeton conference, USIA Deputy Director for Policy and Plans Thomas Chaikin Sorensen, a former newsroom director at KLMS in Lincoln, Nebraska, identified the American "purpose" as one of leading "the technical revolution in communications." The USIA, Sorensen insisted, had mounted an "offensive" in this area, and it intended to "stay on the offensive and to keep the initiative insofar as it is possible and prudent." Sorensen specified that the agency utilized the latest mass media techniques to underscore "the dangers of Soviet and Red Chinese anti-nationalism and neo-imperialism in Africa and Asia," and to provide "positive support of free choice and democratic modernization." Unquestioning faith in America's message and the ability of the USIA to deliver it enabled officials such as Sorensen to stand confident in their ability to "expose the false ideology of Lenin." Such arrogance left the only room for blame on the side of the agency's intended audience, between those "who lack faith in American ideals" and "would keep us on the defensive."[13]

A NEW FRONTIER

Despite the growing unrest in the RVN, which included a near successful coup led by Army of the Republic of Vietnam (ARVN) Colonel Duong Van Dong on November 11, 1960, American support for the Saigon leadership remained firmly in place. John Fitzgerald Kennedy, the American president who took office at the beginning of 1961, possessed a strong track record from his time in the United States Congress as both a staunch anticommunist and a supporter of Ngo Dinh Diem. At the outset of his administration, Kennedy authorized a fifty thousand troop increase for the ARVN, sent in more American advisors, including four hundred Green Berets, ordered the CIA to organize commando raids against the DRV, armed RVN provincial Civil Guard units with heavy weapons, and increased Diem's aid package by forty two million dollars.[14]

As Kennedy steadily increased the American commitment to the RVN, he also set about creating a larger role for the USIA in the nation's foreign policy. During the 1960 presidential campaign in which he had narrowly defeated the Republican candidate and Eisenhower's vice president Richard Milhous Nixon, Kennedy had made American prestige an issue. In the battle for hearts and minds overseas, Kennedy considered the perceived standing of the United States in relation to its communist adversaries as an essential element. While his opponent accused him of "running down America," Kennedy insisted in a series of televised debates that "we have not maintained our position and our prestige." Polls taken by the USIA overseas that leaked to the press after the Eisenhower administration refused to make their results public appeared to support Kennedy's contention. The polls put the Russian approval rating in Great Britain, for example, far ahead of the number for the United States (55% to 12%).[15]

Consequently, Kennedy proceeded to retool the USIA once he entered the White House. In the process, he further blurred whatever distinction remained between informational activities and psychological warfare. Mindful of American prestige, the newly elected president issued a revised USIA Statement of Mission that unashamedly called for the PSYOP program to "help achieve United States foreign policy objectives." Kennedy wanted the agency to concentrate on "influencing public attitudes in other nations."[16] He pushed the PSYOP program to work more fervently to promote American ideals abroad, and to worry less about fostering mutual cultural understanding with other nations.[17] The USIA received increasing responsibilities. And it prospered.

The USIA opened over two dozen posts in Africa, defused reactions to the Bay of Pigs crisis, and heralded the resolution of the Cuban missile crisis. It concentrated much of its efforts on the formerly colonized parts of the globe, areas that officials conceded they had "too long neglected." In Latin America alone between 1962 and 1963, the USIA opened ten new offices and increased its book translation program from one hundred and thirty to two hundred and twenty-five titles, while increasing the number produced in total circulation from a half million to one million copies. It increased its Spanish language broadcasts over short wave radio from two to nine hours daily, and it began a three-hour daily Portuguese broadcast into Brazil.[18]

While it is evident that the USIA vigorously expanded its program into previously overlooked parts of the globe and worked to promote American policy, it is also clear that the people the agency dispatched knew little about the people who resided in the areas where they went. Douglas J. Baker, a former USIA motion pictures officer, recalls his experience photographing the Nile in Khartoum. He decided to include in his shot a group of nomads he spotted at river's edge. At the moment he prepared to film, he felt a tap on his shoulder and "came face to face with the fiercest one of them," who "put his hand on my hands and muttered something in Arabic." When told by his accompanying translator that the group considered it a curse to be photographed, Baker responded with an exclamation of "nonsense." He gave the translator a coin that he claimed to be a "magic amulet" capable of removing all such curses. Baker then "sneaked a shot unbeknownst to him and got into the car."[19]

Baker's experience is hardly an exceptional one. The USIA put personnel into the field, or rotated them among offices, according to perceived agency needs, staff seniority, and technical competency. Ironically, for an agency dedicated to cross-cultural communication, language skills and cultural knowledge of the area did not constitute a prerequisite for assignment. Instead, the USIA relied on the foreign staff it hired to sufficiently breach the cultural chasms its American officers faced. According to Kenneth R. Sparks, former USIA, Chief, European Research Division (1965–1967), specializing in a particular area's language and culture actually constituted an impediment for one's career. He notes that "the top pay" went only to the "broadgauged generalist." The "cultural affairs specialist" found himself "bumping against the top pay bracket in a relatively few years.[20]

Kennedy had Cold War on his mind regarding the USIA, not cultural understanding. He endorsed the agency's effort to "combat the Soviet propaganda trap contained in the phrase "'peaceful coexistence" by using language that reflects the "nature of the conflict in which we are engaged." He called for the agency's "widespread currency and usage" in referring to the conflict as one between the "world of free choice and free cooperation and the world of coercion."[21]

Edward R. Murrow, the renowned World War II reporter and CBS correspondent who Kennedy chose to direct the USIA, echoed the president's tone and pushed his agency to animate it. Murrow called for his agency to "reflect the nuances and special emphases, as well as the main thrust, of the policies and intentions of the U.S. Government."[22] Consequently, the agency also resumed production of hard-line propaganda films. The agency's *Five Cities in June* touted America's achievements in space while sneering at Moscow's exaggerated portrayal of the Russian program. When the USIA screened the film at the United States Mission to the United Nations, the Soviet delegation "left the theater in a huff, understandably irritated over the space commentary."[23]

Kennedy also added the USIA to the Special Group, Counter-Insurgency, that his administration established early in 1962. The Special Group came into existence because of the administration's belief that subversive insurgency represented a "major form of politico-military conflict equal in importance to conventional warfare." In addition to the USIA, the group also included the departments of State and Defense, the CIA, the Joint Chiefs of Staff (JCS), the Agency for International Development (AID), and the White House. It received orders to establish "broad lines of counter-insurgency policy" and to ensure a "coordinated and unified approach to regional or country programs." Kennedy's decision to include the USIA in the Special Group opened the door the agency walked through in Vietnam wearing its new JUSPAO apparel.

A SENSE OF URGENCY

Infused with the anticommunist energies of the Kennedy administration, the USIA received the opportunity to broaden the range of techniques it employed to bring its ostensibly objective perspective to counterinsurgents fighting communists in every corner of the world. The USIA promptly established its own course that dealt with the nature of insurgency movements in other countries. The course focused on the exploitation of insurgents by communists, and it outlined what USIA officials deemed to be America's proper response. The course also stressed the problems presented to the United States by "mass idea—communications" in developing areas. It presented the techniques developed by the USIA for counterinsurgency operations. The information agency offered its staff as lecturers on unconventional methods to the National War College, the Central Intelligence Agency, and the Marine Corps Command and Staff School. It also participated in the work of the interdepartmental committee that Kennedy charged with designing a special five-week course entitled "Problems of Development and Internal Defense." The USIA provided the instructional staff for the special session that took place at the Foreign Service Institute during the summer of 1962.[24]

Indigenous communist uprisings like the one they encountered in Vietnam provided Americans with newfound feelings of both urgency and righteousness about their nation's historical mission. They discerned the inroads made by communists in developing nations as a cry for help from a people whose nation lacked the political will to resist. USIA officials took their cue from political leaders who saw the conflict going on in Vietnam as evidence of a larger communist plan. In the eyes of Americans, Vietnam constituted an "ideal environment for the communist 'master plan' to take over all of Southeast Asia." With regard to the question of Vietnamese self-government, there existed "neither the time available nor any sound justification for 'starting from scratch.'" Officials concluded the presumed external threat posed to the RVN required an immediate shoring up of its defenses. They saw it as America's primary need "to focus the U.S. effort in South Viet-Nam on the immediate internal security problem."[25]

Officials called for Americans "to infuse" Vietnamese "with a sense of urgency and a dedication to the over-all U.S. objective." They saw the conflict as having less to do with the Vietnamese who appeared to be waging it than with America's aim to thwart communist assertiveness. Political leaders mandated for PSYOP "the operational flexibility needed to apply the available U.S. assets in a manner best calculated to achieve our objective in Viet-Nam." They saw it as imperative for American PSYOP "to impress on our friends, the Viet-Namese, and on our foes, the Viet Cong, that come what may, the U.S. intends to win this battle."[26]

Clearly evident in this assessment, the prevailing American view held that the struggle in Vietnam constituted a war of external aggression. Communists owed their success to outside infiltration, organization, terror and duplicity. It is a notion Americans proved reluctant to relinquish despite the robust evidence of southern discontent they witnessed under Diem's oppressive policies. As George McT. Kahin observes, to do so "would have exposed the speciousness of a major rationale for intervention—'aggression from the North'—and also the unpopularity of Diem and subsequent U.S. supported regimes." It also would have undercut the fundamental assumption that Americans clung to regarding their own "objective" perspective on international affairs. It would have forced them, in a way that they either proved unwilling or unable, to question the gap that others detected between their democratic rhetoric and their support for reactionary regimes like Diem's and those who would follow.

The deeper Americans became involved in Vietnam, the more staunchly they cited their difficulties as stemming from the superior tactics used by outsiders to organize and manipulate Vietnamese resistance. Two years after the United States directly intervened in 1965 by sending American military forces to defend the RVN in an ultimately unsuccessful effort, officials continued to cite communist tactics and not indigenous discontent as the root of the problem. "The Communist insurgency in South Vietnam is basically a triumph of organization." American officials believed:

> The Communists do not have any widespread or deep-rooted emotions working for them as they did in Viet Minh days when they were able to capitalize on a pervasive surge of nationalist sentiment and opposition to continued French rule. Nor do the Communists now have any mass base of genuine popular sympathy. What the Communists do have is an efficient, flexible, tightly disciplined and usually locally-rooted organization, carefully built up over a long period of time. It is this organization, which holds the insurgency together and keeps it going.[27]

This view did not allow for the possibility that a demand side of the equation might also exist, that Vietnamese had not been manipulated. Instead, they autonomously had moved to support the side they believed most willing to deliver the independence and socioeconomic reform they had long sought. This American mindset precluded considering this latter scenario plausible.

The idea of being "outperformed" by the communists emerged as particularly strong within the PSYOP program, where officials had long fixated on questions of technique and organization and perpetuated the idea of Asian malleability. Even after he retired from his role of directing the JUSPAO, for example, Barry Zorthian suggested that "it might be argued with considerable validity that the greatest American weakness in Vietnam was organization." He advised that the United States take from its experience in Vietnam lessons that bear an uncanny resemblance to the notions its officials took with them there in the first place. "More than anything else," Zorthian proposed, "the direction of any future United States effort in counterinsurgency needs the distinguishing characteristic of a coordinated, integrated single channel of political/military authority and organization."[28]

SAVING FACE

American PSYOP officials acknowledged early the importance in Vietnamese culture of preserving face "Face is important to the peasant, as to all East Asians." Where "face is lost," the USIA research staff concluded, "there can be no harmony, and harmony is more important than abstract principles of justice."[29] Setting aside for now the stereotypical image psychological operators conjured regarding apathetic Vietnamese, they failed to consider that the same importance might be attached to the idea of "saving face" by American leaders. With each fresh round of criticism leveled against RVN rule from critics of American policy, officials committed themselves more rigidly to making the RVN at least appear like the solution to Vietnam's problems. Worried about America's international prestige, an element they considered vital in their nation's global war against communism, American leaders felt compelled to present to the rest of the world a RVN success story.

The USIA became a primary component of this effort, ultimately reducing its operational role in Vietnam in order to free up resources for the purpose of selling America's policy there to audiences in other countries and to constituents back home. Creating the appearance of hearts and minds won, not actually winning them, is the outcome that came to matter most to PSYOP officials.

American officials had first recognized during the Kennedy administration that, in regard to their efforts to support the RVN "The press problem needs attention."[30] Whether critical of the United States for being too harsh on Saigon, as they had been in the 1950s, or calling for America to push Vietnamese harder for reform, as they did in the 1960s, reporters in Vietnam managed to create constant headaches for officials.[31] By the beginning of the 1960s, officials observed, "The impasse between the RVN and American press corps is well nigh insoluble."[32]

The group of reporters that provided officials with concern consisted of about a dozen Americans who had become highly skeptical of America's unflinching support of the RVN. Its members, individuals such as David Halberstam of the *New York Times* and Pepper Martin of *U.S. News and World Report*, felt constant pressure from Vietnamese officials, who resorted to various forms of harassment, including expulsion from the country, to present RVN policies in the best possible light.

Reporters also chafed under the restrictions placed on them by United States officials. According to Neil Sheehan, one of the reporters then in Saigon, "The confrontation had occurred because of the unprecedented consistency with which we were questioning details."[33] The details Sheehan and the others sought about the conflict often proved not forthcoming from the tightly sealed mouths of United States officials. It remained "the long-standing desire of the United States government to see the American involvement in Viet-Nam is minimized, even represented as something less than in reality it is."[34] Officials failed to grasp this policy's potential to erode American credibility with journalists. In addition, they did not consider what message their actions imparted to RVN officials already prone to enact censorship policies. To no avail, in 1961, USIA Public Affairs Officer (PAO) Wes Anspacher observed:

> We appear to be in danger of getting caught in the web of somewhat conflicting principles. We cannot do more with the image of the RVN than is made available by the RVN itself; in other words, unless we rely on distortion, whatever we mirror will be a reflection of the face the RVN presents to the world. I would imagine that this conviction has prompted USIA's suggestion that we take up the censorship question with the RVN for the umpteenth time. If we were to do this while at the same time we are more or less assuring the RVN that the U.S. will try to "discourage" some elements of the American press, it seems to me we would appear to be talking out of both sides of our mouths at the same time.[35]

In 1963, RVN officials provoked a rebellion on behalf of Vietnam's Buddhist population that brought the issues involving America's press to a head. The Diem government had earned the resentment of Vietnamese Buddhists, a group that claimed to constitute eighty percent of the people under RVN rule, by refusing to revoke Decree #10, which the French issued to identify Buddhism as an organization and not a religion. Buddhists considered it a humiliation that Diem's Catholic dominated regime allowed Catholics to carry papal flags while celebrating the Silver anniversary of Diem's brother, the Archbishop Ngo Dinh Thuc but prohibited them from displaying their own ensigns to commemorate the Buddha's birthday. To protest their inferior status under Diem's rule, Buddhists organized a peaceful march on May 8, 1963, in Hue. RVN Civil Guard opened fire on the assembled group, killing eight and wounding four in the process.[36]

For the American press, the violent reaction taken by the RVN against the Buddhists brought Diem's dictatorial rule into the open and raised questions about United States support for a government so willing to trample such a fundamental American value as freedom of religion. One month later, a seventy-three-year-old monk named Thich Quang Duc sat in a lotus position in the middle of a Saigon intersection. He immersed himself with gasoline then took a lit match to his body to make the ultimate sacrifice as a way to protest against RVN anti-Buddhist policies. The photograph of the incident taken by the Associated Press photographer in Saigon, Malcolm Browne, circulated widely and served as an embarrassment to American officials.

Buddhist activists quickly found that they possessed an ally among the media gathered in Saigon that, to the dismay of embassy officials, stood "predisposed to give full play to the demonstrations, the Buddhist grievances and the government's counter measures." The protesters notified reporters in advance of processions and demonstrations to ensure maximum coverage and to keep the issue alive. The correspondents proved willing to cooperate with organizers as American officials remained largely silent regarding the oppressive tactics utilized by the RVN to handle the crisis. For reporters who, according to the United States embassy, appeared "sympathetic and understanding" of all efforts to keep the crisis alive, an opportunity now existed to drive a wedge between America and the RVN by revealing the full picture to an increasingly attentive and horrified American public.[37]

As the crisis developed, RVN officials made clear to their American counterparts their view of those who covered it for the press. They complained that their government needed "not only to cope with its hot war, but also a continuing Cold War with American correspondents in Saigon." They believed the group of reporters had "succeeded in isolating the Government of Vietnam and then presenting us as monsters to the American people and to the world." Meanwhile, they insisted it would remain RVN policy "that subsequent demonstrations will be put down 'with brutality if necessary.'" United States officials responded by indicating their conviction that "in the end the facts of the situation, as seen by both the Government of Vietnam and the American government, would prevail and these would provide American public opinion with a balanced and accurate picture of the Vietnamese situation."[38]

American officials also promised RVN leaders they would do whatever possible to facilitate painting such a portrait in the media. Robert J. Manning, Assistant Secretary of State for Public Affairs, pledged to "see a number of influential and fair minded editors and to discuss with them candidly the situation in Vietnam." Manning pledged:

> He would convey to them the fact that it was necessary to view developments in Vietnam in the total world context. He would convey to them the need in the complex Vietnam situation for self-discipline on the part of the American correspondents.[39]

Manning noted that, while he could not guarantee the desired results, he did think "it both wise and necessary to make the effort."[40]

American officials decided that the reporters gathered in Saigon represented the "prime determinants of opinion." They also adhered to the idea that the present crisis occurred at the very moment when the United States had at last "turned the corner to defeat Communist subversion." For the moment, at least, they remained committed to the regime's leadership and set about to devise a plan to work on the press. While they considered the acrimonious relationship between RVN officials and reporters as beyond repair, they did believe the Americans could take steps to enhance the image projected of their Vietnamese allies.

Embassy officials commenced a "concerted effort" to obtain the publication in a broad range of periodicals in the United States of "authoritative articles on the situation in Viet-Nam." They pushed stories that stressed "the exciting and unique nature" of America's commitment along with the "Communist penchant for victory by subversion." Regarding the reporters in Vietnam, they called on officials in Saigon to "embark upon a concerted effort to woo" them and to treat them as "politically important individuals, rather than as a group of socially objectionable and professionally incompetent young cubs."[41]

The Buddhist crisis and the lessons American officials took from it stand out as a significant development in the American conceptualization of PSYOP in Vietnam. The controversy made clear for American leaders "the direct and potentially influential role that correspondents on the scene can play in the execution of a foreign policy." Until the incident between Diem and the Buddhists unfolded, embassy officials had expected to receive the unquestioning approval correspondents had afforded United States policy in recent wars. John Mecklin, the United States Public Affairs Officer in Saigon, admitted that the difficulties with the American media in Vietnam constituted "a reality which had been widely overlooked in developing the US advisory role here." This oversight, Mecklin observed, "inevitably complicated US operations in Viet-Nam." He proposed, "How to handle it, hopefully, should be one of the major 'lessons learned' in Vietnam."[42]

The "lessons" officials took included their acknowledgement that the embassy had proven "disingenuously selective in its information, about even minor stories, which might prove unhelpful to the RVN or to the US effort." Its representatives took the acquiescence of the reporters for granted, often treating them with disdain. Their attitude, in combination with the reprehensible behavior of the RVN, left officials in the unprecedented position of facing a "passionately hostile press corps." To fix the problem, officials deemed it expedient to "establish good personal relations between senior

Embassy-MAC/V officers and individual correspondents." They called for personnel to use the "same tact and skill" in a "concerted effort to woo individual reporters."[43]

Henry Cabot Lodge Jr., who replaced Fredrick Nolting as the United States ambassador in Saigon, assumed the point of this operation to win the hearts and minds of reporters. Lodge, a well-known Republican and former political rival of Kennedy, arrived in Vietnam in August 1963, the day after Diem's forces had carried out violent raids on Buddhist pagodas throughout the RVN. Lodge's selection not only provided a bipartisan slant to American policy in Vietnam, it represented a new no-nonsense position regarding Diem.[44] Within days of Lodge's arrival, Mecklin notified the new ambassador that, while it stood as "notably presumptuous" to "discuss press relations with a man of your experience," he felt compelled to do so because "the problem here is not only extraordinarily difficult but also unique."[45]

In general, Mecklin advised Lodge to "invite the newsmen" to "share in the agonizing." That way, he suggested, there existed "at least a chance that you can carry some of them along in your eventual decision." Specifically, Mecklin recommended, "Limit your public comment to an absolute minimum, and always avoid any kind of public judgment about how things are going." He proposed for the ambassador a sort of "divide and conquer strategy," telling him not to be "too easily accessible to correspondents." According to Mecklin, the key for Lodge involved maintaining a "gee-what-should-we-do atmosphere" while working to include "a responsible correspondent in as many social functions as possible."[46] Mecklin did all but label his memo, "How to wine hearts and dine minds." He believed it advanced a viable strategy.

NOTES

1. *The Thirteenth Report of the United States Advisory Commission on Information* (Washington, D.C.: United States Government Printing Office, 1958), 7.
2. *The Thirteenth Report of the United States Advisory Commission on Information*, 9.
3. *The Fifteenth Report of the United States Advisory Commission on Information* (Washington, D.C.: Government Printing Office, 1960), 12.
4. *The Fifteenth Report of the United States Advisory Commission on Information*, 12.
5. Lawrence J. Bassett and Stephen E. Pelz, "The Failed Search For Victory: Vietnam and the Politics of War," in *Kennedy's Quest for Victory: American Foreign Policy, 1961–1963*, ed. Thomas G. Paterson (Oxford: Oxford University Press, 1989), 233.
6. Henderson, *The United States Information Agency*, 23.

7. John Boardman Whitton, "The American Effort Challenged," in *Propaganda and the Cold War: A Princeton University Symposium*, ed. John B. Whitton (Washington, D.C.: Public Affairs Press, 1963), 1–12.

8. George V. Allen, "What the U.S. Information Program Cannot Do," in *Propaganda and the Cold War*, 64.

9. Robert T. Holt, "A New Approach to Political Communication," in *Propaganda and the Cold War*, 45.

10. Allen W. Dulles, "Strengthening American Information Efforts," in *Propaganda and the Cold War*, 73.

11. Holt, "A New Approach," in *Propaganda and the Cold War*, 53.

12. *The Fifteenth Report of the United States Advisory Commission on Information*, 4.

13. Sorensen, "New Directions Under the Kennedy Administration," in *Propaganda and the Cold War*, 68.

14. Bassett and Pelz, "The Failed Search for Victory," in *Kennedy's Quest*, 231.

15. "Popularity v. Power," *Time* magazine November 7, 1960, 12.

16. "Memorandum by the President to the Director, U.S. Information Agency, Describing the Mission of the Agency," in U.S. Congress, House, Committee on International Relations, *Public Diplomacy and the Future, Hearings before the Subcommittee on International Operations of the Committee on International Relations*, 95th Congress, 2nd session (Washington D.C.: U.S. Government Printing Office, 1977), 37.

17. Wieck, *Ignorance Abroad*.

18. Sorensen, "New Directions Under the Kennedy Administration," in *Propaganda and the Cold War*, 71.

19. Henderson, *The United States Information Agency*, 32.

20. Kenneth R. Sparks, "Selling Uncle Sam in the Seventies," in *The Art and Science of Psychological Operations*, I: 118.

21. Memo, An Effective Counter Theme to "Peaceful Coexistence," July 14, 1961, *FRUS: Information Policy*, 127.

22. "Voice of America Policy," quoted in Sorensen, *The Word War*, 239.

23. Green, *Propaganda Abroad*, 36–37.

24. Administration History [2 of 2], I: 53–59.

25. Report, A Program of Action To Prevent Communist Domination of South Vietnam, May 1, 1961, *FRUS 1961–1963, Vietnam, 1961*, II: 8.

26. Report, A Program of Action to Prevent Communist Domination of South Vietnam.

27. Memo, Attacking the Vietnamese Communists' Southern Organization, March 17, 1967, "Memos (A) Volume 67, March 1–19,1967," box 41, NSF—Country File—Vietnam, LBJL.

28. Report, Where do We Go from Here, February 1970, Barry Zorthian papers, LBJL.

29. Memo, to President Lyndon Baines Johnson from USIA director Leonard Marks, January 4, 1966, "USIA Volume 6, December 1,1965 [2 of 2]," box 75, NSF—Agency File USIA, LBJL.

30. Memo, "A Report on South Vietnam," from the Director of the Bureau of Intelligence and Research (Roger Hilsman) and Michael V. Forrestal of the National Security Council Staff to the President (John F. Kennedy), January 25, 1962, *FRUS, Vietnam: January–August 1963*, III: 55.

31. Memo, from the Counselor for Public Affairs of the Embassy in Vietnam (John Mecklin) to United States Ambassador (Henry Cabot Lodge), August 24, 1963, *FRUS*, 621–25.

32. Memo, Memorandum of a Conversation, Department of State, July 26, 1963, *FRUS*.

33. Sheehan, A *Bright and Shining Lie,* 315.

34. Report, Report From the Assistant Secretary of State for Public Affairs (Robert Manning) to the President (Kennedy), undated, *FRUS, 1961–1963,* III: 531.

35. Memo, Memorandum from Public Affairs Officer (Wes Anspacher) to Ambassador in Vietnam (Frederick Nolting Jr.), December 23, 1961, *FRUS, 1961–1963, Vietnam 1961*, I: 339.

36. See Robert J. Topmiller, *The Lotus Unleashed: The Buddhist Peace Movement in South Vietnam, 1964–1966* (Lexington: The University of Kentucky Press, 2002).

37. Report From the Assistant Secretary of State for Public Affairs, undated, *FRUS, 1961–1963,* III: 538.

38. Memo, Memorandum of a Conversation, Saigon, July 17, 1963, *FRUS*, 497–515.

39. Memo, Memorandum of a Conversation, Saigon, July 17, 1963, *FRUS*, 497–515.

40. Memo, Memorandum of a Conversation, Saigon, July 17, 1963, *FRUS*, 497–515.

41. Report From the Assistant Secretary of State for Public Affairs, undated, *FRUS*, 537.

42. Report From the Assistant Secretary of State for Public Affairs, undated, *FRUS*, 534.

43. Report From the Assistant Secretary of State for Public Affairs, undated, *FRUS*, 543.

44. Topmiller, *The Lotus Unleashed*, 3.

45. Memo, "Press Relations," August 24, 1963, *FRUS, 1961–1963*, III: 621.

46. Memo, "Press Relations," August 24, 1963, *FRUS*, 623–24.

Chapter Seven

Picking Up the Slack

OVERVIEW

The first section of this chapter, "Crowning a Communications Czar," chronicles the continued perception of United States officials that the primary problem for their nation's policy in Vietnam stemmed from the unfavorable coverage provided to the RVN by American reporters. In the tumult following the assassinations of RVN President Ngo Dinh Diem and United States President John F. Kennedy in 1963, American officials in the Lyndon Baines Johnson administration struggled to devise a way to manage the situation. As the American commitment to the RVN deepened, so, too, did the mentality that American prestige rested on the outcome of the Vietnamese struggle. Consequently, officials moved to provide the psychological operations program in Vietnam with the singular direction that the USIA proposed that it required to create, at least, the appearance of stability in the RVN.

The second section, "Confrontation Syndrome," illustrates the increasing hostility that existed between the Johnson administration and American reporters who continued to question the viability of RVN rule even as the United States moved toward direct military intervention on the Saigon government's behalf. By the spring of 1965, American officials had come to the conclusion that the war on the battlefield constituted only one part of their conflict in Vietnam. The other part, they deemed, involved the perceptions of reporters who suggested that the hostilities stemmed more from southern discontent than from external communist aggression. As a result, the Johnson administration made the unprecedented decision to create a single umbrella organization to carry out American psychological operations, the JUSPAO. It obliterated any traditional distinctions that remained regarding the conduct of American information activities and psychological warfare.

CROWNING A COMMUNICATIONS CZAR

As American officials set about attempting to build an amicable relationship with the press in Vietnam by courting reporters, the assassinations that occurred in November 1963 and ended the lives of both John F. Kennedy and Ngo Dinh Diem altered the leadership of both the United States and the RVN. Neither crisis did much to change the perception of dim RVN prospects held by the growing number of disaffected journalists. Leaders in Washington characterized the stories of these reporters as "misleading" and suggested they had "exaggerated" unrest "within the South Vietnam government." Officials believed even the most fundamental battlefield reporting remained biased against the RVN. Chief of Staff of the United States Army, General Earle Wheeler, cited a "page one story in the *New York Times*" he identified as "quite misleading." Quoting American advisors, the article alleged the ARVN was ripe with "inertia and low motivation."[1]

Wheeler protested that rather than the bleak scenario depicted by the paper, "Among military advisers there is a growing sense of accomplishment even though the war is heating up." While he conceded that the "Viet Cong forces are tending now to stand and fight harder," Wheeler claimed not enough had been made of the fact that "the Vietnamese are also fighting much better." Contrary to what The *New York Times* contended, Wheeler argued that Americans "should be encouraged by the progress which was being made."[2]

Of course, Americans measured progress in Vietnam in different ways. Only the day before Wheeler called attention to the *Times* article, the USIA reported that the RVN stood "sorely lacking in personnel with the motivation and training to do the job." While the agency proceeded to suggest how it could fill this vacuum it had identified, Wheeler depicted a scenario in Vietnam more conducive to the idea of invading the DRV that he and his fellow Joint Chiefs urged President Johnson to adopt as his own. It is clear from Wheeler's language that he considered the "Viet Cong" as a separate entity from Vietnamese. He hoped to convince Johnson that sufficient stability existed by this point within the RVN to expand the war.[3]

Such pessimistic reporting as Wheeler cited could hardly have been reassuring to an administration headed by a president who felt grave insecurities regarding his new position. Johnson recalled feeling as a "pretender to the throne, an illegal usurper" after taking office following Kennedy's untimely demise.[4] If anything helped ease Johnson's mind about the acceptance of his ascension, it was the efforts made by the USIA to provide a smooth transition in the public's mind. Throughout his career, as a senator and as Kennedy's vice president, Johnson expressed cynicism about the information program. While serving as Majority Leader of the Democratic Party in 1958, for example, Johnson openly lamented that the USIA "does not know what it is doing."[5] As vice president, Johnson often complained that the USIA failed to drum up the enthusiasm that he expected on his official visits abroad.

The USIA changed Johnson's mind about its value by immediately initiating an aggressive campaign on behalf of his presidency following Kennedy's murder. Within a week of Johnson's swearing in ceremony, the USIA had compiled a pamphlet of over eight thousand words about the life of America's new president that it distributed in over twenty different languages. It filmed Johnson's first address as President before Congress and shipped the coverage to one hundred and twenty three points in one hundred and seventeen countries while securing its broadcast on Mexican television the same night of the speech. The information agency also prepared a special briefing kit that covered Johnson's career and made it available to eight thousand American business executives working abroad under "a continuing arrangement with 450 major U.S. companies."[6] Johnson now realized the potential power the agency offered him to shore up his claim to occupy the Oval Office. He especially liked the work of USIA photographer Yoichi Okamo and eventually gave him the assignment of exclusively covering White House activities.[7]

With new confidence in the USIA, Johnson appointed Carl T. Rowan as its director. His choice possessed a mix of journalistic and diplomatic experience, as Rowan had been a prize-winning reporter who entered public life first to serve as Deputy Assistant Secretary of State for Public Affairs and then as America's Ambassador to Finland. In very short order, Johnson pointed Rowan in the direction of Vietnam, dispatching him and Secretary of State Dean Rusk there on a fact-finding mission in April 1964.

In his report to the president, Rowan identified that "the weakest part of the war operation" is located "in the field of information and psychological operations." He urged that Americans pick up the slack from a Vietnamese government that "is sorely lacking in personnel with the motivation and training to do the job required." He noted that:

> In view of the importance of public opinion in Viet-Nam, in this country and in the world at large, I believe that top priority should be given to a large scale United States program to improve the RVN ability to win the support of the people and to tell its story abroad.[8]

Rowan informed the president that he had already met with South Vietnamese officials about a stepped-up effort in psychological operations. They expressed "eagerness to have United States assistance." He concluded that, for America to succeed in Vietnam, the "USIA must be represented on all committees in Washington which are mapping plans to carry us to victory."[9]

While Rowan advocated that America devote more resources to its information program in Saigon, he also believed that it might still fail "unless it has the clear direction of a single individual." Rowan suggested that the task of orchestrating his proposed program be given to Barry Zorthian, "one of the most capable men in this entire agency."[10]

At the time Rowan recommended him to Johnson for the new post, Zorthian served as the agency's Public Affairs Officer (PAO) in Vietnam. Zorthian had arrived in Saigon on February 12, 1964, after doing a stint as a deputy PAO in India. He originally joined the Foreign Service in 1961 when legendary World War II reporter and the USIA director Edward R. Murrow convinced him to leave his position as the Program Manager for the Voice of America (VOA).

In Rowan's view, Zorthian appeared to be the ideal person to oversee the type of comprehensive PSYOP effort he envisioned for the United States in Southeast Asia. He possessed combat experience with the United States Marine Corps during the Second World War and had worked as a reporter before assuming his various roles at the USIA. Zorthian knew how to operate in a war zone, he understood military operations, and he had been well schooled in both journalism and government information policies. Three days after he wrote his memo, Rowan promised the National Security Council that "a great improvement" could be made in Vietnam with "the expenditure of a very small amount of money."[11] He set out to draft a more detailed proposal of his ideas and immediately devoted an increased amount of USIA resources to intensifying its program in Vietnam. Rowan increased USIA staff in Vietnam by adding ten new positions, and he also authorized the direct transfer of personnel from any post in the world to fill the agency's needs in Southeast Asia.

By May, Rowan had initiated VOA programming that targeted people in the DRV for three hours daily. He pledged to increase the Voice's output to over five hours daily by mid-July, and he arranged to boost the station's signal by adding three more fifty-kilowatt short-wave transmitters. Additionally, he worked out an agreement with the Defense Department to carry a jointly run psychological campaign into the countryside that the RVN claimed.[12] He called for his agency's endeavors to be a "built-in part of every program drawn up to meet the crisis in Southeast Asia."[13]

From the beginning, the focus of Rowan's stepped up plans for psychological operations in Vietnam involved the media there covering events. He and other American officials considered Vietnamese opinion primarily as it related to securing popular approval elsewhere for United States support of the RVN. Creating the appearance that Vietnamese accepted RVN rule and not necessarily securing their approval mattered most in the eyes of officials. Image counted most at home as well where Americans grew increasingly concerned with their nation's policies in Vietnam.

Rowan complained at the end of May that "Zorthian and the USIA staff are still not in a position where they have any clear responsibility for dealing with the American press." He found himself "bothered" by the tendency of the American media in Saigon to "emphasize American mistakes and acts that could be called 'brutal' and to give little coverage to Viet Cong atrocities."[14]

Rowan told President Johnson that the USIA probably could not bring to an end the "spate of critical articles written by American newsmen." He did believe, however, that "Zorthian and other USIA officers ought to be free to make whatever effort they can to inform and give guidance to American newsman."[15] Rowan communicated to Johnson:

> USIA has given its highest priority to South Viet-Nam. In terms of its own program and that of the Government of South Viet-Nam, we are moving as rapidly in Viet-Nam as is possible and sparing no facility or personnel in our efforts to see that the rest of the world knows what is at stake in Viet-Nam and Southeast Asia.[16]

The administration provided Rowan with an opportunity to present his ideas on how to improve America's psychological operations in Vietnam at a conference co-chaired by State Secretary Rusk and Secretary of Defense Robert S. McNamara on June 2, 1964, in Honolulu, Hawaii. American officials had gathered to talk about the overall situation in Vietnam, and Rowan's assessment of America's information program in Southeast Asia received much attention. The USIA director made clear the need for operations within Vietnam to achieve an "overlapping effect (e.g. on US newsmen)." "We must make a new start," he proclaimed, "to cure suspicion and cynicism."[17] He recommended that the United States needed to issue a new information directive, mold a more cohesive press program, and minimize military sounding off about the RVN. While the conference participants agreed that they "should not short-change the job inside [the] SVN [RVN]," their focus remained on the United States. McNamara noted that sixty-three percent of Americans continued to be unconcerned about what transpired in Vietnam, but that one-half of the remainder thought the current policy unwise. He recommended that "Zorthian's job be enlarged to handle this job."[18]

After the conference adjourned, Rowan reported to President Johnson that there "was a remarkable degree of unanimity with regard to the steps that ought to be taken." He told the president that those in his administration that had gathered in Honolulu agreed that for the United States to succeed in Vietnam, American officials must demonstrate a "fuller appreciation of the role of propaganda than has heretofore been exhibited."[19] The president's special assistant for national security affairs, McGeorge Bundy, echoed Rowan's view to Johnson. Bundy spoke to several of the officials who attended the conference in Honolulu, including his brother, assistant secretary of state for Southeast Asian affairs, William P. Bundy. Afterwards, he told the president "the one major new agreement growing out of Honolulu is that we need to centralize authority for public information on Vietnam."[20]

Only a few days later, on June 6, 1964, Johnson gave his approval to place one person in charge of America's public relations concerning Vietnam. Following Rowan's recommendation, Johnson named Zorthian to the new position and gave him "overall control of dealing with the American and foreign press in Saigon." Zorthian's responsibility included "jurisdiction over the total information effort." It entailed "a continuing effort to publicize such information as will refute and counteract erroneous and misleading reports now emanating from Vietnam." Johnson hoped Zorthian could accomplish this task by helping "newsmen cover the positive side of the news."[21]

Within a month of the appointment, the United States issued a new set of public affairs policy guidelines. They observed, "Broad American Public and Congressional support are a requisite for the success of United States policy in Vietnam." This directive asserted that the necessary support for American actions would not be attained in the atmosphere of "suspicion and distrust" that results from "denial to the media of full coverage." As a consequence, the new criteria for public relations stipulated that the activities of "all elements of the mission be aimed at maximum candor and disclosure consistent with the requirements of security."[22] Almost inconceivably, officials detected no conflict between their call for "maximum candor" and their creation of Zorthian's new post, one they informally dubbed as "communications czar."

Despite Zorthian's new authority, less than a year later, the USIA chief conceded that "efforts to develop among the South Vietnamese people a sense of national unity and a determination to fight off Communist aggression, have not progressed to my satisfaction." Rowan proposed that he take a trip to Saigon in order to "have another look at the situation there myself."[23] In March 1965, Rowan traveled to Vietnam and met with American and Vietnamese information officials who worked in Saigon and in the provinces. He reported to President Johnson that the USIA "has proved its ability to help the Vietnamese Government meet the insurgency problem." American efforts to open lines of communication with the Vietnamese "have met with success in several local areas."[24]

Still, Rowan perceived a great deal of room for improvement. He continued, "This program is still far below what is required to do the job, and is considerably inferior to what USIA and this country are capable of doing." He added, there could be no excuses for "our failure to mount a psychological effort commensurate with the challenge we face," and he repeated his conviction the United States had "no chance of winning this war unless more and more [Vietnamese] are won over." Rowan's plan for addressing the inadequacy of America's information program in Vietnam necessitated, in his view, overcoming two fundamental problems on the American side. He cited the effort's lack of unified control and direction and its lack of resources as the main impediments to its effectiveness.

Central to his prognosis, Rowan underscored, was the need for Johnson to issue another declaration establishing USIA primacy in the field of informational activities, along with an elevated rank within the United States mission. "With the proper mandate," Rowan insisted, "we will do the job." He emphasized, though, "These programs to affect what the Vietnamese people think must be given the same status, the same concern, the same adequacy of working resources, as our military and economic operations."[25]

Rowan's latest report, and his idea to bring together all aspects of America's psychological operations program in Vietnam, met with a warm reception in Washington. President Johnson's special assistant for national security affairs McGeorge Bundy deemed Rowan's analysis a "good document." Bundy relayed to Johnson that he had run Rowan's proposals through the Departments of State and Defense and found "general support for Rowan's approach and a readiness to encourage USIA in going forward." He believed no time should be wasted in putting Rowan's recommendations into action, and he urged Johnson to capitalize on Rowan's "good report and readiness to act." He drafted a memo for Johnson to sign that would convey a general order for Rowan to "get cracking on his program."[26] Johnson needed little convincing. At a meeting only the day before receiving word from Bundy, he had called for the creation of a program "designed to match and even outmatch the military efforts" underway in Vietnam. The JUSPAO came into being.[27]

CONFRONTATION SYNDROME

President Johnson professed to USIA officials that he had come to embrace the potential of psychological operations in Vietnam. Indeed, he began to show "tremendous interest" in these activities. Johnson even came to suggest his own ideas for future programs to Zorthian directly. He told the JUSPAO director that the key to winning over the support of Vietnamese resided in demonstrating the principle of land ownership. Johnson recommended that the USIA should "take aerial photographs of a man's land and give that peasant a picture of his land that he can hang on his wall that he can see physically."[28]

Such proposed schemes as this one, on one level, demonstrate the disconnectedness of Johnson and other American officials from Vietnamese. Johnson imagined that a farmer, who for generations had been oppressed and kept from owning land by corrupt colonial administrators and landlords, needed to be inspired by the prospect of land ownership. Surely, Vietnamese farmers required no American motivation to seek out land to cultivate as their own. Those individuals who did need such stimulus, the evidence suggests, worked for the United States government.

The "deeply Conservative cast" of the RVN and America's dependence "on that government as an anti-Communist surrogate" led the United States to support initiatives that "largely bypassed such basic questions as power-sharing, social mobility, and distributive justice." American officials ultimately proved willing to offer Vietnamese only "lofty rhetoric and material benefits" in place of "concrete proposals for political reform." This type of support for the RVN that the Americans offered "precluded the kinds of political and social reform that might have allowed a Southern regime to compete successfully with the Viet Cong." While Americans in the field sometimes recognized this dilemma, "policy makers usually did not."[29]

American officials could find rationalization for their views in the work carried out by government-sponsored social scientists from Michigan State University (MSU) such as anthropologist Gerald Hickey. This group, which also included Leonard Finkle and Joseph Zasloff, both political scientists from MSU, identified the village as the basis of Vietnamese politics and society and the place where to focus efforts of political development. Outside of MSU, this village-centered approach to winning Vietnamese hearts and minds met with approval with researchers such as Edward Mitchell of the RAND Corporation and Raymond Gastil of the Hudson Institute.

Throughout the 1960s, this group of scholars offered American officials a clear message. In order to generate greater loyalty to the RVN, local "notabilities" required a boost in status and power. Members of this group contended that poor peasants tended to be more docile and less aspiring than large landowners. Inequality of land tenure, they argued, contributed to greater RVN control by fostering a relationship of necessity between officials in Saigon and wealthy landlords. As historian Jefferson P. Marquis astutely observes, while not explicitly stated, the message the village-centered scholars consistently offered American defenders of the RVN a clear message, "Do nothing to disturb the basis of traditional rural society where it still existed."[30]

In his study regarding the CIA and pacification, historian Thomas Ahern offers that American officials never "treated political or social reforms in

South Vietnam as ends in themselves." Instead, they "got attention only to the degree that they looked like useful tools in building resistance to the Communists." American leaders remained fixed on bolstering the RVN through "a combination of force of arms, US-supplied economic benefits, and propaganda."[31]

Certainly, the notion of how best to sell the RVN, especially to the American people, preoccupied the heart and mind of Johnson the politician. Whenever the president proposed to the JUSPAO a PSYOP scheme, he avidly expressed interest in how such an idea might best be maximized by the agency for positive exposure. Johnson viewed the JUSPAO as a marketing tool for his administration's Vietnam policy. Impressed by the way that the USIA had handled his ascension into the presidency following Kennedy's assassination, he now looked toward the agency to handle public relations for the war in Vietnam as its central task.

According to Zorthian, "psychological operations as such" appeared to the Johnson administration as the "less important side" of the JUSPAO's mission. He explains that Johnson had grown "very perturbed about press coverage, very concerned about the nature of it." The media became an ever-present factor in Johnson's decision-making regarding the war, and he wholly viewed the JUSPAO as a way for him to exercise some sort of control over what reporters said about Vietnam. Zorthian recalls, "The first three or four times I saw the President, I barely had time to say 'yes, sir,' before he took off on the press and just held forth." Johnson "was obviously very frustrated by the coverage." Zorthian remembers him routinely "chewing us [JUSPAO] out, 'Goddamn it, why can't you handle that press out there better? Why do you let them write that? Why don't you straighten them out?' et cetera, et cetera."

Johnson let Zorthian know that he hated journalists scrutinizing his decisions. He "would just fulminate against them, he'd get so goddamned mad." The president would rip the JUSPAO operation "up and down for all the sins of [David] Halberstam and Mal[com] Browne and anyone else he could think of" then turn his wrath toward Zorthian.[32] Johnson "couldn't stand the thought that the press would get some information that he felt was his prerogative to release on his own terms." He called on Zorthian to bring an end to the critical commentary reporters relayed back "to the American public and created opposition to the war."[33] Johnson made it clear to Zorthian he expected nothing less from the JUSPAO than to exercise complete control over the stories that emanated from Vietnam. As historian Waldo Heinrichs demonstrates, Johnson had "no compunction about coloring, shading, evading, and denying the truth, nor deceiving, misleading, punishing, and rewarding reporters."[34]

While Zorthian acknowledges the JUSPAO occasionally "got the lash from Washington," he refuses to recognize claims made by the American press that his agency acquiesced to Johnson's impulse to clamp down on them. In fact, Zorthian refers to the Vietnamese conflict as the "first uncensored war." He points out that the JUSPAO "had very few written restrictions on correspondents." Only tactical information remained off limits to those who covered the war, and "this was agreeable to correspondents. Everything else was a reporter's responsibility. He could write anything he wanted." Under his watch, Zorthian contends, "it was very rare that there were any obstacles to communications."[35] It is evident, nevertheless, that Johnson's shadow constantly lurked over the operation and in large measure determined JUSPAO priorities. Over the course of one particular flap with the media, Rowan wrote to Zorthian:

> In any event, I am emphasizing that the President is deeply interested in this because he is concerned about anything that threatens to erode his public support back here. I think you will understand this.[36]

In his agency's defense, Zorthian contends that reporters had become infused with "confrontation syndrome." He complains that his agency constantly found its work mired in criticism over the public relations role Johnson had assigned it to play in the war. American journalists derided the agency's daily news briefings as the "Five O'clock Follies."[37] While American officials had come to consider psychological operations as a total information and public relations effort that included all means of communication, reporters saw the situation differently. They resented their government's attempt to manage their product by utilizing the same tool it employed to conduct what they labeled as psychological warfare. Rather than engendering the good publicity for its policies the Johnson administration hoped for, the JUSPAO created the impression that the American government had something to hide from its own public. It thereby heightened the already palpable level of tension that existed between government officials and the media.

George Beebe, president of the Associated Press Managing Editors Association, loudly complained that the "information given newsmen about the United States' military role in Viet Nam is under supervision of the United States Information Service (USIS)." He questioned how the press in Vietnam could do its job with its access to information about the war limited to what it received from an agency that only aimed to "present the image of the United States in the most favorable light." Beebe accused the Johnson administration of using the USIA to "propagandize the American public." He cited the penchant the USIA demonstrated for "requiring escorts" when reporters interviewed soldiers and for the "controls" it exerted over military briefings.

Time magazine's Southeast Asia Bureau Chief Frank McCulloch concurred. He called USIA-generated material "shaded" and often "incomplete." Meanwhile, Associated Press reporter Bob Eunson alleged that the JUSPAO called journalists out on the carpet whenever they wrote stories critical of America's effort. He believed Barry Zorthian, in particular, "privately criticized newsmen for stories that had nothing to do with national security."[38] USIA Director Rowan responded to the criticism directed at his agency by staunchly defending Zorthian, and by contending "censorship is impossible under present circumstances in Viet Nam." He proposed, "Censorship would only work if it was handled jointly with the Vietnamese Government, and we feel this would raise vastly greater problems than it would solve."[39]

Critics of the USIA remained unconvinced by Rowan's argument that the JUSPAO did not engage in censorship. Some went on to accuse his subordinates of not only censorship, but also of fabrication. In an article he penned for the Associated Press entitled "'Battle' is Staged for USIA," Peter Arnett accused the USIA of deliberately confusing fact with fiction to generate support for American policies. Stationed in Saigon, Arnett traveled south to investigate reports of a battle in Tan Hiep and found a different sort of action than what he had expected. He discovered the USIA filming an attack scene for a 30-minute film it sought to distribute worldwide. Arnett quoted a USIA spokesperson, who stated that the film's purpose is to "show the world how things really are here." The goal was "realism."[40]

Arnett questioned the USIA commitment to portray what he and other reporters saw as the reality of Vietnam by pointing out that, during its film shoot, the agency had "avoided the real war and such major engagements as Binh Gia."[41] His story implied, not very subtly, that the USIA had every intention of releasing an obvious piece of what government officials would otherwise label propaganda under the cover of a documentary account of the war. Arnett accused the USIA of making a "people-to-people film" that built a case for American involvement in Vietnam by minimizing the conflict's brutality, and he included commentary critical of the USIA effort made by American soldiers. One United States Army captain he interviewed advised him that if the USIA wanted to portray the war, "they should try to do it at least like it really is. All this colored smoke and gung ho bayonet charges are for the birds." The American soldier told Arnett that, while he thought his mission to Vietnam would involve fighting the enemy, "They might as well have sent me to Hollywood."[42]

Rowan answered the charges made by Arnett by suggesting "what was phony was not the movie, but the AP story." He labeled the report a "product of Arnett's imagination." Still, the USIA director conceded to his agency's contracting of a civilian camera team that worked to "re-stage some of the violence and terror tactics of the Viet Cong" in order to "present an accurate picture of the dirty war South Viet Nam is forced to take on." He argued the USIA only used the technique because "no camera team could know where the Viet Cong might strike on a given night and have its cameras in position so as to film the actual event."[43]

For the sake of projecting what USIA officials deemed an accurate portrayal of the war, they demonstrated no compunction about using whatever tactics they viewed as necessary. They detected no difference between filming the version of the war they scripted and the one that actually took place. Nevertheless, in order "to protect USIA's credibility," Rowan agreed publicly to order that "no staged footage was to be used" in the movie.[44] American reporters were not contented with Rowan's effort to appease them. They continued to criticize the agency's efforts to manufacture an administration-friendly view of the war for mass consumption. While President Johnson had created the JUSPAO to help newsmen cover the positive side of the news, the very notion that their government engaged in psychological operations to steer the direction of their reports begged their skepticism.

Contrary to its intended effect, the very existence of the JUSPAO compounded America's public relations problems in Vietnam, and it helped spawn opposition to United States policies, particularly from American journalists. The JUSPAO blurred the line between psychological operations in Vietnam and media accounts of the war in the United States and in the rest of the world. American officials remained oblivious, however, of the glaring contradiction created by their nation's espousal of a free press and the actions they took to shape the stories written by reporters. They further perceived no problems contained in subverting the very democratic ideals they sought to impart to the Vietnamese as they went about their own work. Even the USIA Library in Saigon, purportedly an institution to educate Vietnamese about democratic freedoms, came under fire for its treatment of "books critical of United States policy," which it "neither listed in its card catalogue nor displayed on the shelves."[45]

Officials had come to interpret the truth about the war narrowly so it only consisted of whatever particular angle of a story that had official approval at the time. They considered themselves engaged through the JUSPAO in prosecuting a vigorous program to assure that the true story of America's war effort safely made it to the United States and elsewhere. Zorthian explained during the conflict:

> I find nothing unusual in our practices here, since our press attaches in every country in the world work with American correspondents as well as with foreign

> journalists. I try to meet the potential problem here by describing my job as two-hatted—one hat being Director of JUSPAO, the other hat being Minister-Counselor of Information in behalf of the Ambassador.[46]

USIA officials never questioned the monopoly on veracity claimed by their agency, and they vigorously condemned as slanted and untrue all views advanced by the media that proved inconsistent with their own. They adopted an "us versus them" mindset in dealing with American journalists in particular. From their vantage point, the Johnson administration was victimized by reporters who appeared to them as bent on subverting the truth in their stories about Vietnam.

Information agency officials reported to President Johnson that America had "suffered seriously from wire photos distributed by American news services depicting atrocities by South Viet Nam and U.S. troops." They did not, however, advise Johnson to impress on American and RVN military leaders the psychological implications of poorly behaved troops. Instead, they complained of the "bias" demonstrated by America's media, and they "repeatedly offered other photographs for transmission" instead. "Finally," the USIA reported that agency "efforts seem to be paying off" after securing the publication of two photographs that it suggested "illustrate the humanity of our effort."[47]

NOTES

1. Peter Grose, "Score Vietnam Inertia," *The New York Times*, April 20, 1964, 1.
2. Summary Record of the 528th Meeting of the National Security Council, Washington, April 22, 1964, *FRUS, 1964–1968,* I: 259.
3. See Kahin, *Intervention*, 210
4. Quoted in Doris Kearns, *Lyndon Johnson and the American Dream* (New York: Harper & Row, 1977), 177.
5. Quoted in Sorensen, *The Word War*, 95.
6. Memo, USIA Weekly Report, December 3, 1964, "USIA," box 135, Confidential File—Agency Reports, LBJL.
7. Green, *Propaganda Abroad,* 39.
8. Memo, Information-Psychological Warfare Program in South Viet-Nam, April 21, 1964, *FRUS, 1964–1968*, I: 254–55.
9. Memo, Information-Psychological Warfare Program in South Viet-Nam, April 21, 1964.
10. Memo, Information-Psychological Warfare Program in South Viet-Nam, April 21, 1964.
11. "Summary Record of the 528th Meeting of the National Security Council, April 22, 1964, *FRUS, 1964–1968*, I: 259.

12. Memorandum from the Director of the United States Information Agency to the President, May 26, 1964, *FRUS, 1964–1968*, I: 382–83.

13. Memorandum from the Director of the United States Information Agency to the President, June 4, 1964, *FRUS, 1964–1968*, I: 451.

14. Memorandum from the Director of the United States Information Agency to the President, May 26, 1964, *FRUS, 1964–1968*, I: 384.

15. Memorandum from the Director of the United States Information Agency to the President, May 26, 1964.

16. Memorandum from the Director of the United States Information Agency to the President, May 26, 1964.

17. Summary Record of Meetings, Honolulu, June 2, 1964, *FRUS, 1964–1968*, I: 428–30.

18. Summary Record of Meetings, Honolulu, June 2, 1964.

19. Memorandum from the Director of the United States Information Agency to the President, June 4, 1964, *FRUS, 1964–1968*, I: 451.

20. Memorandum from the President's Special Assistant for National Security Affairs to the President, June 3, 1964, *FRUS, 1964–1968*, I: 441.

21. Telegram from the Department of State to the Embassy in Vietnam, June 6, 1964, *FRUS, 1964–1968*, I: 471.

22. Memo, Public Affairs Policy Guidance for Vietnam, July 7, 1964, "Material on JUSPAO," box 4A39, Papers of Barry Zorthian, LBJL.

23. Memorandum from the Director of the United States Information Agency to President Johnson, February 24, 1965, *FRUS, 1964–1968*, II: 363.

24. Memorandum for the President, March 16, 1965, "Rowan Report," box 190, NSF—Country File—Vietnam, LBJL.

25. Memorandum for the President, March 16, 1965, "Rowan Report."

26. Memorandum for the President, March 17, 1965, "National Security Action Memorandum (NSAM) 325," box 6, NSF—NSAMS, LBJL.

27. Memorandum by the President's Special Assistant for National Security Affairs, March 16, 1965, *FRUS, 1964–1968*, II: 446.

28. Transcript, Barry Zorthian interviewed by Paige E. Mulhollan, May 6, 1969, 79–45:33, Oral History Collection, LBJL.

29. Ahern, *CIA and Rural Pacification*, 403.

30. Marquis, "The Other Warriors," 85.

31. Ahern, *CIA and Rural Pacification*, 404.

32. Transcript, Barry Zorthian interviewed by Ted Gittinger, May 26, 1982, 84–64: 40–41, Oral History Collection, LBJL.

33. Transcript, Barry Zorthian interviewed by Michael L. Gillette, April 24, 1981, 84–63: 9, Oral History Collection, LBJL.

34. Waldo Heinrichs, "Lyndon B. Johnson: Change and Continuity," in *Lyndon Johnson Confronts the World: American Foreign Policy, 1963–1968*, ed. Warren I. Cohen and Nacy Bernkopf Tucker (Cambridge: Cambridge University Press, 1994), 20.

35. Memorandum for the President, March 16, 1965, "Rowan Report," box 190, NSF—Country File—Vietnam, LBJL.

36. Memo, to Zorthian from Rowan, June 23, 1965, "Material on JUSPAO," box 4A39, Papers of Barry Zorthian, LBJL.

37. Memorandum for the President, March 16, 1965, "Rowan Report," box 190, NSF—Country File—Vietnam, LBJL.

38. Article, "Reporting Facts of Life in Viet Nam," May 7, 1965, "USIA volume 4, April 14, 1965 [2 of 2]," box 74, NSF—USIA File, LBJL.

39. Memo, Memo for the President, May 12, 1965, "USIA volume 4, April 14, 1965 [2 of 2]."

40. Article, "Battle is Staged for USIS," January 13, 1965, "USIA volume 4, April 14, 1965 [2 of 2]."

41. On December 28, 1964, the 9th NLF Division seized the Catholic village of Binh Gia, forty miles east of Saigon. The battle left two elite units of the Army of the Republic of Vietnam (ARVN) destroyed, the 33rd Ranger Battalion and the 4th Marine Battalion. See Phillip B. Davidson, *Vietnam At War: The History: 1946–1975* (New York and Oxford: Oxford University Press, 1988), 324.

42. Article, "Battle is Staged."

43. Memo, Memo For the President, May 12, 1965.

44. Memo, Memo For the President, May 12, 1965.

45. R. W. Apple, "Saigon U.S.I.A. Hides Critical Books," *New York Times*, November 13, 1966, 2.

46. Letter, JUSPAO Director Zorthian to USIA Director Leonard Marks, September 2, 1966, "Material on JUSPAO," box 4A39, Papers of Barry Zorthian, LBJL.

47. Report, Weekly Report for the President, February 14, 1967, "USIA—1967," box 135, Confidential File—Agency Reports—USIA, Papers of Barry Zorthian, LBJL.

Chapter Eight

Dwelling on Difference

OVERVIEW

This chapter contains three sections. The first, "These Poor People," illustrates the ahistorical and racialist view regarding Vietnamese that the JUSPAO propagated through its efforts. American officials constructed an image of Vietnamese according to an oriental ideal that situated them in not only another place, but also in a time other than the contemporary world. They suggested that Vietnam constituted the product of a static culture, and its people stood wholly unprepared to act independently against the outside forces that intervened from the modern world. The second section, "They Bend Like Bamboo," argues that Americans believed they understood Vietnamese culture and possessed an understanding of how to win hearts and minds. They moved to introduce a sense of national identity and the components of modernization that they presumed the Vietnamese lacked and must want. They trusted their ability to elicit the response they expected, in part because of their confidence in American technical expertise, and in part because of their assumptions regarding Vietnamese malleability.

The chapter's final section, "Out of Tune," demonstrates the rigidity of American thinking regarding the Vietnamese. Most Americans refused to reconsider their assumptions about the people they encountered. Few bothered to learn anything about the history, culture, or language of a group they ostensibly set out to persuade. Psychological operators have suggested since the end of the war that American policy in Vietnam proved to be a bankrupt one, but the means the JUSPAO employed to support United States aims proved equally hollow.

THESE POOR PEOPLE

United States officials, both in and out of the USIA organization, perceived the Vietnamese as one-dimensional, oriental cutouts that engaged in a game they did not initiate and could not independently conclude. They found the culprits they blamed for setting Vietnamese pieces in motion nearly everywhere except Vietnam. Comments made by United States ambassador Henry Cabot Lodge reveal a prevalently held view among Americans that encouraged them to consider Vietnamese people as inherently passive, with little capacity, political or otherwise, to control their environment. According to Lodge, prior to the appearance of communist "terrorists" in Vietnam, "the lives of these people were peaceful." A sense of widespread "chaos and confusion" among the formerly serene Vietnamese, he observed, had directly derived from the communist tactic of "chasing these unhappy and pitiful people from pillar to post."[1] Considering that as the nation's ambassador Lodge constituted the face of America as well as its top official in Vietnam, a statement that wholly neglected nearly one hundred years of Vietnamese resistance against French colonialism must have astounded any reasonable person within earshot.

In private, Lodge demonstrated even greater candor assessing the situation. In a meeting with American President Johnson and his advisors, he contended, "There is no tradition of a national government in Saigon. There are no roots in the country. Not until there is tranquility can you have any stability." He made it clear that Vietnamese leadership offered little chance of achieving the constancy he advocated its population required. Lodge went on to advise Johnson and his advisors, "I don't think we ought to take this government seriously. There is no one who can do anything." As the United States stood on the brink of pouring troops into Vietnam, Lodge offered justification for the Johnson team to act unilaterally. "As we move ahead on a new phase," he proposed, "it gives us the right and duty to do certain things with or without the government's approval."[2] Lodge minimized the capacity of Vietnamese to govern themselves. Still, the president did not object to his uniformed characterization. At a similar meeting held a couple of months earlier, Johnson himself worried aloud about what he perceived as a lack of initiative among Vietnamese. "I think we ought to put extra money into the USIA," Johnson had observed. He then added: "[President Dwight David] Eisenhower thought we ought to spend much more than we are—we have to give these people a will to fight and a will to win."[3]

Johnson referred to a visit Eisenhower had made to the White House on February 17, 1965, to meet with the president and his advisors about Vietnam. On that occasion, Eisenhower emphasized that "morale was the key factor in

Vietnam." He stressed the need to secure "the participation, and hence the morale, of the entire population." He observed, the "population of South Viet-Nam wants to be friendly to us [the United States]," but it desperately required "information and inspiration." Eisenhower made a strong plea for "proper support for the Information Service." He proposed that the "radio is an outstanding way to motivate." The Vietnamese in the countryside and in the hamlets, he contended, need to know "America is with them." Johnson responded to Eisenhower's recommendations by expressing "much interest." He stated his intention to see what could be done to strengthen these activities in the near future.[4]

Even as Americans doubted the ability of the Vietnamese to rebuff communists independently, they expressed continued skepticism regarding Vietnam's capacity for democratic self-government. In 1961, officials in Saigon reported, "The ideal of a two-party system with a constitutional opposition should be our long-range goal for Viet-Nam." They emphasized their "considerable reservation as to the possibility of accomplishing this in the near future or of the practicality and desirability of actively pressing for it at this time." Officials called for continued American support of the RVN while concurrently warning, "The Vietnamese people lack the necessary sophistication and understanding, as well as the necessary sense of political responsibility to make a two-party democratic system work at this time." The best America could do in Vietnam, officials believed, was to induce the RVN "to adopt, as many as practicable under the circumstances, liberal procedures and reforms which will plant the seeds of democracy and eventually create a solid enough base on which to build still further democratic institutions."[5] Even if Americans could prod the Vietnamese to demonstrate some sort of democratic initiative, they could not trust them to act on it responsibly.

Mark Philip Bradley ably demonstrates that this perception that Americans maintained of Vietnamese people grew out of their inherited turn of the century suppositions about white people and the superiority of Western civilizations. While by the 1960s, Bradley shows, Americans had begun to substitute modernization theory for racialized cultural hierarchies to view the world around them, the resulting perceptions they held about the non-Western world remained roughly the same. They saw the non-West as a developmental project in every sense of the phrase. Americans looked at the under-developed world as unequal to the society they had constructed, and they believed its people had little to offer them as they paved the way for continued global progress. Bradley's work suggests Americans had historically viewed Vietnamese as "politically immature" and as exceptionally "susceptible to foreign influence."[6] They saw Vietnamese society as static, stuck in a bygone era. The JUSPAO program did little to alter this view. Instead, it subscribed to the prevalent view that Vietnam required an American-led tutorial on how to progress as a nation. With enough guidance, officials believed, Vietnam would enjoy economic prosperity and democratic reforms. Moreover, they widely expected that the Vietnamese would recognize the wisdom of America's intervention and express their unwavering allegiance to the United States as gratitude.[7]

At first glance, it appears not only that Americans in Vietnam proved mindful of Vietnamese culture, but it also looks like they developed a deeply seated appreciation of its nuances. The United States information program revealed the potential to enable United States officials to construct a more complicated depiction of Vietnamese culture through its production of booklets such as those entitled "Montagnard Tribal Groups of the Republic of Vietnam" and "The Religions of Vietnam." This literature delves deeper into Vietnamese life than the usual us/them model typically allows. The religion pamphlet, for example, not only discerns differences and commonalties among Buddhist sects such as Mahayana and Theravada, it also points out to what extent practice and ideals diverge in everyday life.[8]

The JUSPAO suggested in its press kit that Vietnam constituted a "mosaic," where "the closer the scrutiny, the more diverse the pattern." It defined different relationships in Vietnam's society to encompass "military vis-à-vis civilian, urban vis-à-vis rural, more conservative politically vis-à-vis those termed 'more revolutionary'; and, among religious groups, particularly the Buddhists, Catholics, Hoa Hao and Cao Dai." It further commented that Americans should not automatically characterize Vietnamese within these groups, as "a sizable number of these differences are tempered by personal factors."[9]

American PSYOP also encouraged steps to make sure individuals in the field executed policy with local culture and individual variation in mind. United States soldiers received wallet-sized cards that Military Assistance Command required them to carry at all times. The list instructed soldiers "Join with the people! Understand their life, use phrases from their language and honor their customs and laws."[10] The orders provided to advisors headed into villages included a heavy emphasis on the need for cultural sensitivity. Individuals assigned to work directly with Vietnamese counterparts obtained training that advised them on the importance of ancestors. They were also taught to listen to what their Vietnamese colleagues had to say and not just issue orders.[11]

Planners even demonstrated their recognition of the need to generate products "tailored to local audiences and developed with full RVN participation."[12] From June 1965 through July 1966, JUSPAO assistance led to the creation of thirty-three provincial newspapers. The agency encouraged the local staffs to work with minimal direction from Saigon. "A province paper is just that," JUSPAO officials contended. It "should not be centrally handled." The agency professed a reluctance to implement an American-style program that operated in the mode of Vietnam's "culture be damned."[13] Information officials also showed concern with overstepping the bounds of their advisory role to the RVN's Vietnamese Information Service (VIS). They cautioned against "JUSPAO replacing VIS as the major contributor to information activities." Americans, JUSPAO officials proclaimed, "cannot and should not" operate "independent of the Vietnamese." Anything less than a joint effort, they noted, left them "wide open to criticism and charges of nonsupport."[14]

The JUSPAO underscored that the American presence in Vietnam "is for the Vietnamese a temporary, not permanent, necessity." It cautioned that any action the United States took that projected a permanent American presence risked violating the "four aspects of the Vietnamese psychology":

- Their pride in the ancient cultural civilizations of Indochina;
- Their memories of other foreign presence;
- Their knowledge that Vietnamese carry the main burden of both dying in war and in building their nation;
- Their awareness that the other non-communist peoples of Asia are free masters of their own nations.

The JUSPAO explained that Americans in Vietnam "need to exercise the reflex of achieving an objective not by doing it themselves, but by working through the Vietnamese." The measure for any American's success, the agency reasoned, "is the degree to which he works himself out of a job."[15]

Yet Americans did proceed to operate largely independent of Vietnamese despite possessing an apparent awareness of the hazards entailed by following this course. They did so, not because they minimized Vietnamese culture, but because officials willingly attributed RVN behavior to the cultural characteristics they generally ascribed to Vietnamese. Officials made little effort to discern the difference between the Vietnamese in general and particular individuals such as Diem, who selfishly acted to further their own positions of power. Information personnel routinely cited RVN deficiencies, but they believed Vietnamese, regardless of religion, dialect, or regional origin, could do no better.

American planners advised patience in dealing with the RVN, warning that "Working through Vietnamese officials will sometimes mean achieving progress not in a continuous curve, but in pulses."[16] In other words, they expected little. It is telling that JUSPAO guidance called on American advisors to work "through" rather than "with" their Vietnamese counterparts. Officials presumed the initiative resided exclusively with Americans. They worried about the Vietnamese willingness to keep pace and maintain the appearance of the independent government that Americans sought to cultivate. Harry S. Casler, JUSPAO assistant director, complained, "The fact is we depend on Vietnamese sources and they do not produce—purposely."[17]

THEY BEND LIKE BAMBOO

USIA operators surmised that the considerable number of people who inhabited Vietnam's countryside possessed little interest in national affairs, lacked the ability to grasp abstract political ideals, had no sense of Vietnamese identity, and would line up behind whatever government appeared most legitimate. The inheritance of Lansdale's approach is clear. USIA officials accepted his working proposition in the Philippines then in Vietnam that nationalism in Asian countries could be brought into existence simply by applying the correct techniques. Consequently, American PSYOP officials learned of Vietnamese culture from individuals who conceived themselves to share little in common with the people they governed and whose administration remained out of touch with popular concerns. Because of their virulent anticommunism and the inclination toward simplification that accompanied it, United States officials confused the RVN with the rest of Vietnam that it claimed to rule.

Had Americans proven more able to question the credibility of their alleged allies, for example, they would have discovered the prominent role played by heroes in the history of popular Vietnamese nationalism. In 39 C.E., two sisters, Trung Trac and Trung Nhi, jointly known as Hai Ba Trung, defeated an army of Chinese with a force of peasants under their direction. They became national heroines. In 1284, Tran Hung Dao engaged an army of Mongols in a fighting retreat with his much smaller force. Vietnamese paid homage to his example by tattooing nationalist slogans on their arms in defiance of the invaders and by continuing the practice well into the 20th century.[18] Yet, when the JUSPAO raised the idea of publicizing Vietnamese historical heroes to promote nationalism, the agency faced resistance from RVN officials who were "very frank in their conversations as to why they did not want the 'hero' program. It is not in their culture."[19]

The program ultimately came into existence, but only because of the effort made by United States Army Brigadier General John Fritz Freund, who, acting as director of the Office of Field Services, JUSPAO, began such a program on his own initiative. Freund "was told" by other agency officials about the RVN's attitude regarding the program, but he "rejected it." The general criticized his colleagues as "desk-bound" and insisted a hero campaign would resonate with the Vietnamese outside of Saigon to whom he spoke about the idea.[20] Freund began the campaign, without official JUSPAO support, near the end of 1965. By the beginning of the next year, the campaign's success compelled the JUSPAO to implement the hero concept in television broadcasts nationwide.

Americans developed an understanding of Vietnamese culture by accepting folk sayings relayed to them by RVN officials and anthropologists as truisms. They did little to learn from the actual people they encountered. Instead, they approached Vietnam as if it had been populated with human manifestations of ideal Confucian constructs and not a people looking for a modern path that might lead them to self-government and societal cohesion. They saw Vietnam as a collection of tribal societies that lacked any historical sense of nationhood. Through their racialist lenses, they identified passivity as the only trait that Vietnamese shared with one another. The JUSPAO reported, "The Vietnamese are a merry people and love to laugh."[21]

The JUSPAO drew its personnel from the ranks of the USIA, an agency staffed by communications specialists. To meet the demands placed on it by the increasing American presence in Vietnam, the USIA had officers from around the world "rushed to Saigon with little preparation."[22] Few in the agency knew anything about the Vietnamese, and less still possessed the training to acquire any knowledge. The JUSPAO had a "dearth of people holding advanced academic degrees."[23] Consequently, the agency "turned to the tools of social science and to social scientists themselves for assistance." The JUSPAO research office supervised the work of the Center for Vietnamese Studies of Saigon University and funded its operating costs of slightly over one hundred thousand dollars per year.[24] It published "JUSPAO Guidance" for American advisors on such subjects as "actions to take when faced with civil demonstrations, VC terrorism, and others."[25]

The JUSPAO also drew from the work of American scholars who had turned their attention toward Southeast Asia after the Second World War. The Pacific Relations Institute took the lead in this endeavor, encouraging a number of studies during the 1950s.[26] The JUSPAO looked to this scholarship for insight, and it found simplistic portrayals of Vietnamese that fed already existing stereotypes. Former JUSPAO official Robert W. Chandler, for example, cites Ellen Hammer's work as authoritative backing for his notion that Vietnamese would rather "bend like the bamboo" than risk disrupting social harmony. He quotes Hammer's observation that "certain rules of courtesy" dictate that "the Vietnamese often adapts his answers to what he believes to be in the desires of his interlocutor."[27]

Other scholars that the JUSPAO consulted depicted the Vietnamese as not only passive, but also as childlike. Marilyn W. Hoskins, a social anthropologist who conducted fieldwork in Vietnam, produced a report for the JUSPAO entitled "Building Rapport with the Vietnamese." In it, she depicts the people she studied as a happy and simple group that time had somehow left behind:

> Vietnamese laugh and giggle not only when they think some joke is funny but when happy, when embarrassed, or when covering up emotions they do not wish to show. A Vietnamese enjoying himself, excited over good news, being corrected, describing a horrible accident, or being treated for a painful wound may emit the same laugh.[28]

With the aid of such experts as Hoskins, Americans idealized Vietnamese in such ways as to magnify cultural difference beyond all reasonable bounds. Americans conceived the Vietnamese as inhabiting not only some other place, but also some other time that they considered as far more simplistic than the contemporary world they occupied. It became easy to marginalize Vietnamese concerns.

United States Special Assistant for Vietnam Affairs George Carver, for example, contended that Vietnamese could not be considered "politicized in our sense of the term." Carver expressed the widespread view among Americans that typical Vietnamese expressed little interest in matters extending beyond one's family. This belief held that a national consciousness simply did not exist in Vietnam. It encouraged Americans to see Vietnamese as objects, "to be secured, protected, or controlled" for the sole purpose of defeating communism.[29]

Americans who sought to win Vietnamese hearts and minds considered themselves as ones who brought the means for progress to a backward people. They saw themselves as carrying out their nation's historical mission of manifest destiny. United States Army Lieutenant General William Pelham Yarborough, Assistant Deputy Chief of Staff for Military Operations (Special Operations (1965–66)), observed:

> Perhaps not since the earliest days in the development of the United States had American soldiers been assigned missions that involved bridging such significant cultural gaps. George Washington, Lewis and Clark, Daniel Boone and other American pioneers knew a great deal about Indian customs and dialects.[30]

Yarborough spoke proudly of the work carried out by United States Army Special. Forces and JUSPAO among the Montagnard, the largest of the ethnically non-Vietnamese groups that occupied over half of RVN governed territory, largely the interior highland areas north of Saigon. In 1966, Americans built or sponsored over three hundred schools and provided medical care for numerous communities. According to Yarborough, "The psychological impact of Special Forces among the tribal peoples of Indochina was strong and lasting." PSYOP officials "hand picked" personnel according to their possession of "political sensitivity, tact, diplomacy, inner discipline and initiative." As a result, Americans "often did the right thing by instinct in the absence of specific training."[31]

Yarborough's claims of success are dubious, however, considering that Americans worked to deprive Montagnards of both the cultural identity and regional autonomy that they managed to retain even under French colonial rule.[32] The hospitals and schools that Americans offered them could not undo for the Montagnards the forcible regrouping of their communities that Diem's government initiated in 1959. By the 1970s, despite the hearts and minds that Yarborough believed Americans had won, the Montagnards turned against the RVN, citing Saigon's longtime pattern of exploitation against them.[33]

According to the view that the JUSPAO advanced, "the best and proper behavior" for the people in Vietnam, regardless of their ethnicity, meant knowing and understanding the world and then "adjusting intelligently to it." It advised, "To a large degree this means learning how to accept the blows of a capricious fate with decorum." The American information program contended Vietnamese culture prohibited action and encouraged unlimited flexibility; it called for passive acceptance as opposed to shows of initiative. "The wise man does not try to change his universe, he adapts to it. Bend as does the bamboo in the wind, say the Vietnamese."[34] Americans perceived Vietnam as an unchanging civilization where uninterrupted order, regardless of the cost exacted on society, is all that mattered to people. The JUSPAO reported that, for Vietnamese engaged in conflict, "a compromise acceptable to both sides is the ideal solution."[35]

Throughout America's involvement on behalf of the RVN, officials presumed Vietnamese held little interest and even less capability for contributing to the development of their society. They came to see Vietnamese, as well as others in parts of the world they deemed as developing, as people who possessed "limited concerns and awarenesses." Techniques that information officials had earlier utilized with "more established" audiences they now saw as inapplicable for addressing the "more rudimentary societies of the developing countries."[36] The USIA advanced the idea that Asians especially paid little heed to events beyond their immediate vicinity. They remained "apathetic and withdrawn" while attributing all worldly affairs to divine, not human, will. "Contrary to the basic sense of individual determinism commonly accepted in the West," the agency reported, "the Asian lives in a universe determined by the will of God, not man."

The USIA noted that throughout Asia the view prevailed that "the only hope of peace of soul is not to improve this world, but to escape it." People normally opted "often for passive acceptance of external phenomena." It observed, "This frequent attitude of passivity and reliance upon providence often makes it difficult to help Asians help themselves." Officials interpreted America's need to win popular support for its agenda as a mission to pacify a population it viewed as pre-modern. They believed the United States had entered a land where a "belief in the efficacy of supernatural force is common, xenophobia is characteristic, kinship and stratification promote particularistic loyalties, lack of technological skills and illiteracy prevail, and internal communications remain inadequate."[37] They assumed a combination of force and material improvements would do the job. Time after time they initiated programs that emphasized security and civic action. These programs counterproductively often served to reinforce the worst oppressive tendencies exhibited by the RVN. If Americans built enough schools and hospitals and armed enough villagers, they expected the communists would go back to wherever they had originally come from. The Vietnamese would be left in peace under American tutelage.

OUT OF TUNE

The American PSYOP program worked in opposition to the theories scholars offered at the time based on earlier attempts at international persuasion. Citing his study of World War II, the renowned sociologist Daniel Lerner offers, "Even under the extreme hostility of total war some measure of international persuasion remains feasible." Two criteria, however, must be met by the communicator under all conditions for psychological campaigns to enjoy any type of success. First, it must establish credibility by consistently supporting words with actions. Second, all persuasive efforts must aim to reinforce a target audience's desired behavior rather than to convert it from unwanted actions.[38]

In other words, as suggested by United States Army Colonel Phillip Paul Katz, who served as a psychological operation official, a communicator must harmonize its message with its receiver's current needs. "Being in tune with the audience requires people, and people who understand communication techniques, attitudes, and other people who may be different from themselves." He concludes:

> To gauge the effect of PSYOP programs, it is necessary to consider the objectives and performance of political, social, or economic programs being sup-

> ported. If the primary programs are seriously deficient and these deficiencies are understood by the people, no amount of PSYOP will persuade the people that the programs are worth while.[39]

Americans assumed Vietnamese, many of whom saw the NLF as liberators, the RVN as corrupt, and American behavior as suspicious, would follow their lead when prodded to do so. Evidence suggests Americans repeatedly carried out their work oblivious to the feedback they received from Vietnamese regarding current conditions. They attempted to change, rather than accommodate, the fundamental perceptions Vietnamese held regarding the war. Americans christened the NLF as the Viet Cong, labeled as revolutionaries the reactionary RVN leaders they supported, and presented American interests as wholly altruistic despite the evidence that accumulated over the course of the war that challenged this view of events.

Americans sought to win Vietnamese hearts and minds, without first gaining an understanding of what feelings and ideas they already contained. They never made a sincere effort to connect with the people that they sought to reach. One JUSPAO province advisor, for example, reported from Bau Me Thout that the province faced "serious problems, and the people are quite unhappy." He observed that "the RVN has failed to create the image of a sociopolitically reformist regime." The people in the province "are not receptive to RVN propaganda." Instead, he argued, "People are concerned about the cost of living. There is widespread non-VC opposition to the RVN."[40]

United States leaders erred by casting their lot with a dubious governing apparatus. They then compounded the problem by stubbornly refusing to recognize their mistake. They relied on the RVN, a profoundly unpopular government outside of urban Vietnamese Francophile and Catholic circles, to gauge the state of Vietnam's collective psyche. The problem went unrecognized. As a result, Americans who headed into the field to serve as advisors received instruction about the modern Vietnamese love affair with French culture.[41]

Americans believed, "In Southeast Asia, the ability to speak French is a tremendous advantage; it establishes an important common bond between the adviser and his counterpart as well as other civilian and military contacts."[42] Considering the brutality of the French colonial regime, and the bloody rebellion waged by Vietnamese to rid their nation of France, Americans communicating with their RVN colleagues in French played right into the hands of communist strategists who warned of United States imperial ambitions. It also did not help that so few Americans ever appeared in Vietnam who proved conversant in any dialect of Vietnamese. By 1970, more than fifteen years after the USIA began operating in Vietnam, only six of the one hundred and thirty-two Americans the JUSPAO employed spoke functional Vietnamese.[43]

As a result, even those Americans who spoke Vietnamese could not be sure the message they composed would remotely possess the resonance they ascribed it. In an interview he gave after the war, it is apparent United States Army Lieutenant Colonel James V. Wells, who spent a year in Vietnam with A-Company, 8th PsyOp Battalion, Nha Trang, retrospectively recognized the problem. "You can't have Americans with American ideas," he offers, "putting out messages to Vietnamese who have a different language, have a different culture, have a different way of thinking." Wells indicates one of his officers who "spoke fluent Vietnamese" engaged in "creating propaganda because he understood the words, the language, but he was not sure what dialect he was speaking in or—whether what he was saying was really appropriate to the thinking of the Vietnamese."[44]

According to Harry Latimore, who served as JUSPAO Chief of Plans, Policy and Research, no more than fifteen percent of USIA officers overall and only about thirty-five percent of those in the field possessed any training in Vietnamese whatsoever. Those who did had passed through at least part of the 42-week course offered by the Foreign Service Institute. Even those who completed the entire period of instruction, though, as Latimer acknowledges, "were not fully fluent." Military officers received even less training than did the civilians the USIA employed. They generally received the opportunity to sit through a twelve-week course that provided them with a "rudimentary ability" to understand the language but left them lacking the tools to speak it functionally.

Like other officials, Latimer minimizes any difficulties created by the inability of Americans to communicate with the people who they set out to persuade in the language they spoke. He even suggests, apparently as oblivious as his colleagues to the ammunition Americans supplied their critics who charged them of colonial ambitions, the inability of Americans to speak Vietnamese came to be offset by those who "had an acceptable capability in French." Americans could converse with Vietnamese "whose French was often rusty but who nonetheless were reasonably proficient in that language." Latimer insists USIA officers "were accustomed to frequent dealings with foreign nationals from their previous and normal professional experience, and they fitted easily into the role of advisor, friend and confidant of their Vietnamese counterparts."[45] It eludes him entirely that by speaking any language other than Vietnamese, but in particular by using French, Americans risked unwanted parallels between themselves and Vietnam's former foreign oppressors. The situation did not escape Vietnamese who opposed RVN rule as their PSYOP regularly described officials in Saigon as "lackeys of Americans."[46]

Based on a sampling of the oral history accounts provided by the 25th Infantry Division and published by Eric M. Bergerud, it is reasonable to suggest that the unwillingness of Americans to learn how to speak Vietnamese won them neither the friends nor the confidants Latimer believes they did. "A majority of us couldn't even say hello or good morning in Vietnamese," recalls former member of the 25th Dan Vandenberg. He continues, "The only words we picked up were a mixture of French and Vietnamese, and you couldn't use any of them in polite company."

Vandenberg identifies the Vietnamese expression most frequently utilized by Americans as *ditty mau*, or "Get the hell out of here." This suggestion is hardly a phrase that comes to mind when one aims to make new friends. One of Vandenberg's fellow soldiers adds, "A few GIs really got involved with the Vietnamese and tried to learn their language and customs, but most of us really didn't want anything to do with them." Echoing Vandenberg, he adds, "The only Vietnamese words we learned were the cusswords."[47]

Americans bothered to learn little of the language of the people they supposedly courted because they considered their task more about eradicating communists than about winning the allegiance of Vietnamese, who they viewed more as cultural artifacts than as modern people. They obviously set themselves up for a great deal of frustration as they found themselves unable to cope with the reality of independently minded Vietnamese who sought explanations they could understand for what transpired around them in their own country. American behavior unwittingly enabled the people they encountered to draw the same kinds of simplistic conclusions about them as they had developed regarding Vietnamese. In one survey, villagers stereotyped Americans as "childish and kind" based on their "friendliness toward children." Vietnamese complained, though, that they could not "understand the Americans" because they could not "communicate with them."[48]

Left to judge Americans only by their actions, the Vietnamese suggested that they conduct themselves with greater maturity. In another survey, respondents called on Americans to "refrain from impolite and at times bullying attitudes," in particular, "drunkenness and misconduct toward women." In addition, Vietnamese expressed their distrust of American intentions. They "most frequently stated" reasons for the ongoing war as "colonialism and ideological and economic expansionism."[49] Moreover, province reports identified that the Vietnamese "fear that, when 'negotiation time' comes around, Americans will abandon Viet-nam 'as they did to Laos.'"[50]

Stereotypes became all too common on both sides in Vietnam, as outside of formal USIA programming, Americans made little effort to connect with the people who they had come to save from communism. Even many of the individuals who worked for the USIA did not consider psychology as something they needed to address by the way they conducted themselves. Few Americans realized winning hearts and minds could not be divorced from the manner by which they conducted themselves at all times. Wesley R. Fishel, a political science professor at Michigan State University who advised RVN officials under the auspices of a CIA program, recognized the shortsightedness of this approach. He expressed his recognition that Americans needed to become more conscious of the people whose nation they sought to build.

In 1964, Fishel wrote an article that appeared in *The Washington Post* that raised serious questions regarding the relationship that had evolved over the past decade between Americans and their Vietnamese allies. In the piece, Fishel wondered why after ten years of "intimate contact" Americans had not managed to cross the "barriers of language and culture" that prevented them from understanding Vietnamese "patterns of thought and action." After having spent more than half of the preceding decade in Vietnam, Fishel offered a dismal assessment about the scene he now watched unfold. "In many ways," he observed, "we are no closer to understanding the Vietnamese than we were in 1954." Despite "the most awesome assemblage of 'brass' the United States has dispatched to any single country," Fishel lamented, a consensus had developed "on all sides in Vietnam that programs carefully thought through rarely seem to be carried out in logical and satisfying fashion." According to the professor, "The war is not being won and we are on what looks to be a sharply downward spiral of defeatism and disintegration."

Fishel questioned the "rigid and too often insensitive approach to the problems of Viet-Nam" Americans undertook, and he wondered whether the methods they utilized truly suited the needs of the people who resided in the nation they sought to build. He criticized American mission chiefs who sat "stiffly in their offices" or rode around Saigon "formally in their big cars." Even Fishel, though, could not shed the paternalism he criticized. He suggested that the Vietnamese most needed a "corps of understanding friends" who they could solicit for advice, not a "big brother" that only sought to order them around. "After ten years," Fishel argued, Americans had still not come to grips with their nation's role as the "great and powerful patron" that can "give money and supplies to our tiny ally but cannot command—only offer advice."[51]

Only a few months after Fishel's observations appeared in the newspaper, the USIA director echoed his refrain directly to the White House. Carl Rowan suggested that Americans make "a conscious and concerted effort to develop relationships of warmth, trust and easy collaboration with opposite numbers on the Vietnamese side." He informed President Johnson that his "first finger is pointed" at his own agency. Several USIA personnel, Rowan reported, "have never asked a Vietnamese in for dinner." He continued, "I have made it clear that under these circumstances we will never get RVN officials and the Vietnamese people to take the steps needed to beat off Communist aggression, and that I expect more of USIA officers than this."

Rowan pointed to the general trend of Americans to segregate themselves from Vietnamese by establishing American-only officers clubs and by declaring certain bars in areas like Da Nang off limits to Vietnamese. "From the standpoint of our refuting the persistent VC clamor that 'the Americans are imperialists,'" he contended, "these developments are not good." It is clear, nevertheless, that while Rowan understood better than most Americans that PSYOP consisted of a total effort, his concern chiefly resided with appearances and not reality. He had less interest in fostering a genuine partnership between Americans and Vietnamese than he did of creating the illusion of one's existence. He noted:

> It would be a fine stroke of public relations if we asked top RVN military people to join in the first briefing session—indeed to give their assessment of how the war is going in various areas. We Americans could then hold whatever private sessions we deem necessary—without creating the impression that we meet, make the decisions and then drag the RVN along.[52]

From the American perspective, most of the people who lived in Vietnam appeared as incidental to events that swirled around them. In most cases, Americans saw Vietnamese as people who got in the way of what the United States had set out to accomplish. American assessments at the village level typically complained of uncooperative counterparts on the RVN side who served no purpose other than to slow things down. American advisors were quick to characterize their counterparts as "very slow and short of initiative" or even as "short of good will or capacity." Americans used these types of phrases often in the reports they issued, indicating their expectation of Vietnamese apathy. For example, in one abridged set of province reports from July 1966, the terms "lazy," "apathetic," "fence sitting," and "disinterested" are used over a half-dozen times by American personnel to describe local Vietnamese.[53] USIA representatives in the field considered their Vietnamese colleagues specifically as "very poor about technique."[54] In return, officials in Washington made it clear they understood. "The Vietnamese Information Service cadre, as usual, stink. There has never been any dispute of this point."[55] In the process of such exchanges, they communicated more efficiently with Vietnamese than even they thought possible.

NOTES

1. Article, "Lodge Cites 2,000 Murders By Viet Cong Last Year," August 10, 1965, USIA Wireless File, reel no. 67, Records of the Office of the President and Independent Agencies (REOPIA), LBJL.

2. Notes of meeting, July 21, 1965, "July 21–27, Meetings on Vietnam," box 1, Meetings Notes File, LBJL.

3. Notes of meeting, May 16, 1965, "Meeting with Foreign Policy Advisors.

4. Memorandum of Meeting with the President, February 17, 1965.

5. Telegram, Embassy in Vietnam to the Department of State, January 31, 1961, *FRUS, 1961–1963, Vietnam, 1961*, II: 50.

6. Bradley, *Imagining Vietnam and America*.

7. Michael E. Latham, *Modernization as Ideology: American Social Science and "Nation Building" in the Kennedy Era* (Chapel Hill: University of North Carolina Press, 2000).

8. Pamphlet, "The Religions of Vietnam," April 1967.

9. Booklet, *JUSPAO Vietnam*, January 1966, "Vietnam Kit," box 426, Office Files of Fred Panzer, LBJL.

10. Article, "GIs get rules on how to win Vietnamese," undated, USIA Wireless File, reel 39, REOPIA, LBJL.

11. Handwritten notes, "Relation with Counterpart," undated, "CI File," Lee Papers, MHI.

12. Cable, USIA to Saigon, February 3, 1967, "Psywar," NSF, Files of Robert W. Komer, box 8, LBJL.

13. Memo, Harry S. Casler to Barry Zorthian, July 24, 1966, "Material on JUSPAO," Papers of Barry Zorthian, box 4A39, LBJL.

14. Memo, Robert F. Delaney to Daniel Oleksiw, Undated, "Psychological Warfare," NSF, Komer—Leonhart File, box 21, LBJL.

15. Booklet, *JUSPAO Vietnam.*

16. Booklet, *JUSPAO Vietnam*.

17. Memo, Harry S. Casler to Barry Zorthian, July 24, 1966.

18. Vadas, *Cultures in Conflict*, 60–61.

19. Memo, Harry S. Casler to Barry Zorthian, July 24, 1966.

20. Memo, Comments Upon Completion of Assignment with JUSPAO, June 17, 1966, "Material on JUSPAO," box 4A39, Papers of Barry Zorthian, LBJL.

21. Article, "USIA Profile: Culture and Drama in Vietnam," *USIA Correspondent,* March 1967, "USIA Correspondent 1965–1967," box 14, Papers of Leonard Marks, LBJL.

22. Sorensen, *The Word War*, 284.

23. Chandler, *War of Ideas*, 241.

24. Latimer, *US Psychological Operations in Vietnam*, 18.

25. Report, Psychological Operations Organization and Activities in RVN, April 1966, "US Army Special Warfare School, 1967," Lee Papers, MHI.

26. J. H. Brimmel, *Communism in Southeast Asia* (London: Royal Institute of International Affairs, 1958); Bernard Fall, *The Viet-Minh Regime: Government and Administration in the Democratic Republic of Vietnam* (New York: Secretariat of the Institute of Pacific Relations, 1956); Ellen Hammer, *The Struggle for Indochina, 1940–1955* (Stanford: Stanford University Press, 1955).

27. Chandler, *War of Ideas,* 241.

28. Marilyn Hoskins, "Building Rapport With The Vietnamese," in *The Art and Science of Psychological Operations,* I: 287.

29. Ahern, *CIA and Pacification*, 399.

30. William P. Yarborough, "Special Forces Psychological Operations in Southeast Asia," 338.

31. Yarborough, "Special Forces Psychological Operations in Southeast Asia," 338.

32. Kahin, *Intervention*, 99.

33. Davidson, *Vietnam At War*, 751.

34. Memo, JUSPAO Field Memorandum Number 27.

35. Memo, for President Lyndon Baines Johnson from USIA director Leonard Marks, January 4, 1966, "USIA Volume 6 1 December 1965 [2 of 2]," box 75, NSF—Agency File USIA, LBJL.

36. Report, Administration History: USIA Volume I (2 volumes), 1968, "Volume I: Administration History [1 of 2]," box 1, Administration History, LBJL.

37. USIA report, "Attitudes, Communications and Communist Propaganda: Factors in Insurgency in Southeast Asia," in *The Art and Science of Psychological Operations,* II: 553–70.

38. Daniel Lerner, "Is International Persuasion Sociologically Feasible?" in *The Art and Science of Psychological Operations*, I: 47–48.

39. Phillip P. Katz, "PSYOP and Communication Theory," in *The Art and Science of Psychological Operations*, I: 37.

40. Report, Special Reports to Ambassador Porter, August 19, 1966, "Vietnam Provincial Reports," box 230, NSF—Country File—Vietnam, LBJL.

41. Handwritten notes, Relation with Counterpart, May 12, 1966, "United States Army Special Warfare School—1966," Lee Papers, MHI.

42. Charles C. Thebaud, "Language and the Adviser," in *The Art and Science of Psychological Operations*, I: 312–13.

43. Chandler, *War of Ideas*, 240.

44. Transcript, Lieutenant Colonel James V. Wells interviewed by Colonel Gerald Lord, 1984, Senior Officer Oral History Program, MHI.

45. Latimer, *US Psychological Operations*, 27–29.

46. Report, 1st AIR CAVALRY DIVISION/AN KHE, September 20, 1966, "Revolutionary Development, 1966–68," Lee Papers, MHI.

47. Eric M. Bergerud, *Red Thunder, Tropic Lightning: The World of a Combat Division in Vietnam* (Boulder: Westview Press, Inc., 1993), 222–23.

48. Airgram, American embassy Saigon to Department of State, PROVINCIAL REPORTING: Political Developments in III Corps, June 2, 1967, "Vietnam Provincial Reports," box 230, NSF-Country File Vietnam, LBJL.

49. Report, Nationwide Hamlet Survey (Second Interim Summary Report—II Corps), 1 November 1967, "NSF—Country File—Vietnam," box 59, Revolutionary Development Program 2, March–December 1967, LBJL

50. Report, BINH-THUAN & NINH-THUAN/PROVINCES, September 20, 1966, "Revolutionary Development, 1966–68," Lee Papers, MHI.

51. Wesley R. Fishel, "We Just Naturally Upstage Vietnamese," *The Washington Post*, December 27, 1964, E4.

52. Memo, for the President (Johnson) from USIA director (Rowan), March 18, 1965, "Rowan Report," box 190, NSF—Country File—Vietnam, LBJL.

53. Report, Special Reports to Ambassador Porter, August 19, 1966.

54. Report, Report of the Administration Conference of 2nd Tactical Zone at Dalat, August 11, 1966, "Revolutionary Development, 1966–68," Lee Papers, MHI.

55. Memo, For Robert Komer from Richard Holbrooke, August 19, 1966, "Vietnam Provincial Reports," box 230, NSF—Country File—Vietnam, LBJL.

Chapter Nine

Toward Oblivion

OVERVIEW

This chapter is comprised of five sections. The first, "Propaganda of the Word," examines the programs the JUSPAO sought to implement in the Vietnamese countryside. It notes the perceptions of officers in the field who often blamed American frustrations on Vietnamese colleagues who they deemed as unmotivated. While some American advisors in the countryside distinguished between apathetic RVN officials and Vietnamese generally while pointing out popular discontent with the regime the United States supported, officials in Saigon and Washington exhibited little interest. They continued to see the primary problem facing the United States as an indifferent Vietnamese people.

The chapter's second section, "Tootbrushes and Television Sets," illustrates the superficial way American PSYOP went about attempting to secure Vietnamese hearts and minds in the countryside while doing little to respond to demands for real reform. It consisted of a superficial "bread and circus" routine to provide onlookers of tangible evidence that the RVN possessed a connection to the people it ostensibly governed.

"Smoothing Ruffled Feathers," this chapter's third section, demonstrates the inability of Americans, despite the "integrated approach" of the JUSPAO, to identify the interconnectedness of the conventional war with the battle for hearts and minds. People dislocated by the military's effort to isolate communists often received little more than token gestures from United States officials concerning their situation. The chapter's next section, "The Other War," illustrates that, despite the rhetorical commitment made to communicating with the Vietnamese, America's leadership demonstrated little interest in providing the JUSPAO with either the resources or the time it required to pursue its assigned mission. "Kill Them All," the chapter's last section, suggests that military officials considered PSYOP in a way congruent to their institutional history. From commanders to soldiers "in country," the American force considered PSYOP to possess only the tangential benefit of providing an occasional means to secure a few defectors.

PROPAGANDA OF THE WORD

Little quarreling took place over the deficiencies Americans routinely identified as existent in the RVN, but neither did much scrutiny. Most ascribed the condition to Vietnamese nature. At least a few Americans working in Vietnam's provinces, however, detected patterns indicative of other reasons for the situation. One advisor "thought long to determine why the cadre do not want to help their own people, especially since they are getting paid for it." He ultimately concluded, "They do not feel that they are working for the Vietnamese but for the Americans; therefore, the cadre feel that they do not have to work at all." The advisor went on to suggest that, if his rationale proved correct, provincial programming "is leading to a state of oblivion."[1]

Another province representative noted American-Vietnamese cooperation tended to hinge on the extent that Vietnamese perceived their interests as compatible with those of their partners. Too often, he explained, the priorities of Vietnamese and American counterparts "are at a greater or lesser degree of variance, a factor which mitigates against optimum coordination of effort."[2] Yet another one reported confirmation that "when RVN and US interests coincide, obstacles to cooperativeness and effectiveness melt away."[3]

The dominant American view held that all the Vietnamese really want is peace and quiet, regardless of the government that ran the country. The key to pacification, many believed, amounted to "the capacity to pacify, *i.e.*, to beat the hell out of the guerillas and *thereby* convince the peasants that the VC is a loser." While Americans rhetorically committed to winning Vietnamese support, the opinion prevailed among them that Vietnam's population showed little inclination and even less ability to engage in ideological debate. Military force and not psychological persuasion was the instrument they identified as the key for achieving a satisfactory resolution.

After returning from Vietnam one official advised, "Like others in the world, peasants love a winner and are much too smart to pick a winner by reading one of Zorthian's seven billion leaflets."[4] American had long reached the conclusion, based on their views of Vietnamese as apathetic and communism as an external threat, that contentment resulted from security. In their minds, the reverse did not appear conceivable.

One report filed by a provincial advisor in 1966 openly questions whether the agency's effort truly coincided with what local Vietnamese sought. The official describes media output as "well executed" and its "materials well written and well organized." He notes, however, "They have relatively little impact." The report cites the problem in this particular case as resulting from "propaganda of the word rather than the deed," but continues, "Even where there is a 'deed,' people do not necessarily get enthused." The document goes on to recall a dedication of two refugee classrooms in Cau Ca hamlet outside Duc Hoa. It describes "people whose attendance had been commanded" and who "were just going through the motions." It offers:

> There was no spontaneity to their actions. They were obviously tired. So, is the answer more schools, or more schools (and other projects) that the people really want? Is anyone really interested in finding out what the people want? Apparently not. Here was a fine new school that drew only yawns.[5]

The question raised by the advisor went unanswered. The United States continued to push programs that brought only tepid acceptance from either the RVN or other Vietnamese. Chieu Hoi or, as Americans called it, Open Arms, borrowed heavily from similar American initiatives in places like Korea and the Philippines. Implemented nationwide but targeted at "communist-infested areas," Chieu Hoi offered "the hand of friendship to those who wish to return to the GVN." The program sought to induce defections from "military and civilian Viet Cong (VC) and members of the North Vietnamese Army (NVA)." It was administered by the JUSPAO in a combined effort with MACV, USAID, and the Vietnamese Chieu Hoi Central in Saigon.

The JUSPAO publicized the program through every available means, from loudspeaker to leaflet. It distributed "Safe Conduct" passes that promised each potential returnee (Quy Chanh) fair treatment. The program offered vocational training and financial assistance to resettle once a returnee completed a maximum of forty-five days of "motivational training" in one of the provincial Chieu Hoi centers. The highest ranking defectors were taken to the national Chieu Hoi center at Thi Nghe, near Saigon, where the JUSPAO attempted to exploit them for "Psywar puposes."[6]

From its earliest stages of development, the program proved to be a difficult sell in Vietnam. United States officials noted "Vietnamese reluctance to embark on a meaningful amnesty program." They reported in 1963 that the RVN had "finally developed a plan, but it is far from satisfactory." The "basic problem," they identified, "is revealed by the Vietnamese insistence that what they want is not an 'amnesty' policy but a 'surrender' policy."[7] Outside of RVN officials, Vietnamese never warmed to Chieu Hoi either. After promoting the program's benefits for four years, Americans reported they had yet to convince average citizens in the RVN that Chieu Hoi was not "an R&R program for Viet Cong." Vietnamese commonly complained it was "not right to give Quy Chanh better treatment" when there were "other poor people in Vietnam who never were VC at all."[8]

RVN authorities resisted the idea of reincorporating their former enemies back into the fold. Chieu Hoi ran throughout the war, but it never managed to achieve the level of success Americans hoped it might reach. According to an analysis of the program conducted in 1967, Chieu Hoi suffered from "administrative constipation." RVN leadership had cooperated with the United States in running the program but staffed it with "largely inept VIS civil servants." While the study cited many problems that plagued the operation, it concluded, "The heart of the matter, however, is that the RVN is still not fully sold on the program across the board."[9] That point is evident from even briefly scanning the feedback provided from the field regarding the defector inducement program.

The JUSPAO advisor in Ba Xuyen noted, "On the surface, the Chieu Hoi Program in the province seems to move along. Ralliers do come in, are processed and leave." He went on to indicate, though, that all did not work as well as it appeared, "But any look below the surface reveals grave problems." The program lacked coordination with other elements, experienced labor difficulties, and reflected its RVN chief, whose "character is marked by aloofness and a desire not to cooperate."[10]

In Hau Nghia, the JUSPAO reported, "The Chieu-Hoi office has yet to work out a re-indoctrination program of the most elementary sort." The assessment alleged "ralliers still just sit at the center with only magazines to read." Any attempt to facilitate the reintegration of defectors into society, it pointed out, "is shamefully lacking."[11] Reports filed from Kien Phong observed "very little if any activity" took place there, despite the office's recent increase in Vietnamese personnel. JUSPAO advisors indicated they were "unable to find" any evidence that Kien Phong operation had, in fact, made any new hires at all.[12]

Assistant Director for Chieu Hoi Ogden Williams reported to Washington that the Chieu Hoi program at the province level was "very weak." He called the vocational centers for Quy Chanh "even weaker." Williams described an operation riddled with holes made by Vietnamese who did not subscribe to its mission. He pointed to "would-be returnees" that did not defect because they were "afraid they'll get no job after leaving the centers." Williams could not confirm or deny the validity of that fear because, "At present, no one in Vietnam can tell you what happens to QCs [Quy Chanh] after they quit the center." He lamented that once a Quy Chanh left a Chieu Hoi center, "The man mostly just walks off down the street."[13]

Clearly, Vietnamese resistance to the program did not diminish from when USIA director Carl Rowan had assessed it over a year earlier. In early 1965, Rowan visited Chieu Hoi rehabilitation centers to find them "presently without even the most elementary psychological indoctrination programs." He observed that returnees "languish two or three months in 'detention' with few daily activities beyond interrogation and occasional lectures on the 'good national cause.'" As a result of such conditions, "Defectors have become disillusioned anew, returned to the VC fold, and are now warning their colleagues not to fall for the promises of the government's Chieu Hoi program."[14]

In most cases, the RVN merely went through the motions in carrying out a Chieu Hoi program it possessed little interest in pursuing. But in some instances, its reception of the operation could be downright hostile. The JUSPAO reported from Lam Dong, "Little change. Despite a large scale appeal, few ralliers came in." On an even more "sour note," it pointed out that the local Chieu Hoi Center "was razed by a RVN battalion stationed across the road." The marauding troops "tore part of the walls from the building and stole tables and beds to be used for fire wood." Subsequently, "a strong protest was made to the battalion commander," but it received no response.[15]

Despite the difficulties they encountered from the RVN in trying to implement Chieu Hoi, American officials continually attempted to reinvigorate the program. The results were consistently disappointing. RVN officials complained Americans took too strong of a lead in planning and carrying out programs that ostensibly originated with Vietnamese. They complained of JUSPAO failures to prescreen "certain Chieu Hoi films, and that there were certain errors in them which should be corrected."[16] Little had changed in the process from a year earlier, when RVN officials had pointed to the leaflet program that they believed had "become too closely identified with the Americans."[17]

Assistant Director for Chieu Hoi Ogden Williams referred to his position as "close to the most thankless job I have ever been connected with." He described his staff as "somewhat bloody and bludgeoned" from trying to gain acceptance for a program that "presents difficult practical and psychological problems for the Vietnamese and will take a little longer." Williams remained optimistic, though, thanks to the budget support and "sense of priority" Chieu Hoi received from officials in Washington.[18] American leaders acknowledged the "lagging" state of Chieu Hoi. They recognized its shortcoming resulted from "lukewarm ARVN attitudes, and poor staffing at provincial levels."[19]

American officials continued to prioritize Chieu Hoi despite Vietnamese concerns because of the program's potential to gain support elsewhere for

United States policy. Former JUSPAO director Barry Zorthian identified Chieu Hoi as one of the "benchmarks in Vietnam" that the United States could cite as a measure of its progress.[20] Officials in Washington had "long regarded the defector program as major asset." They considered it to show "more effectiveness per dollar" than "any other GVN activity" in producing a "significant psychological effect on public opinion US and elsewhere."[21] Moreover, this view of the Chieu Hoi program's value did not only prevail among officials in the United States. Officials in Saigon conducted the program to court an audience outside of Vietnam. Assistant Director for Chieu Hoi Ogden Williams initiated a special campaign to extend voting privileges to Quy Chanh with "largely international opinion in mind."[22]

The American concern to gain acceptance outside of Vietnam diverted from the United States effort to win hearts and minds among the Vietnamese. The reports that JUSPAO advisors filed throughout Vietnam routinely point to Vietnamese dissatisfaction with the RVN as well as to their unease with American actions. One account notes that "inflation continues to be a corrosive element in RVN-citizen relations." It continues by identifying persistent economic difficulties that "prompt sharp questions from the public regarding the ability or desire of RVN to deal with major problems." The same report suggests the speed at which RVN opposition moved to exploit such opportunities. It observes that Vietnamese doubts "are reinforced by a sudden increase of VC activity," such as an incident at Ke Sach that "scored an effective psychological point for the VC."

Another memo from Hau Nghia observes, "Real interest in serving the people instead of merely executing the minimum program requirements laid down by Saigon remains low." The report asserts that Vietnamese perceived RVN officials as "all too dedicated to 'being' instead of 'doing,'" and suggests, "Nothing short of their early retirement will ease the problem." Meanwhile, from An Giang Province, the JUSPAO advised that people stood "very much aware of the election for the Constituent Assembly." Their enthusiasm remained tempered, nevertheless, "due to the fact that they feel as if they have realized nothing from the promises of candidates who were elected in past elections."[23]

Popular frustration also mounted with RVN armed forces operating in the area. According to the report, "Incidents involving ARVN [Army Republic of Vietnam] and civilians and the quartering of troops on the rural population are bringing about expressions of dissatisfaction from the people in the Long Xuyen area." The JUSPAO representative in Hau Nghia cited the occurrence of "shoot-outs" between ARVN regional forces on two separate occasions. One transpired in Cu-Chi and the other on the main street of the province capital "in front of the police station, the province chief's house and the USAID house." Such incidents only "increased the disaffection between the forces and the people."[24]

In addition to making complaints regarding ARVN behavior, many Vietnamese questioned the fundamental ability of the RVN to protect them. According to a description of events in Quang Tin, Saigon's forces provided them with little relief. It notes that, after insurgents burned "scores of families" from their homes, "one of the camp leaders—asked if he'd like to have better protection from RVN forces—said 'Ha! Send me Koreans! Never mind RVN troops.'" The JUSPAO representative responsible for Binh Duong and Phuoc Long reported, "RVN prestige is suffering as a result of its inability to provide adequate protection for the citizens."

Evidence is abundant in the provincial reports that American advisors often held their RVN colleagues in low esteem because of behavior they viewed as evidence of poor motivation. The feedback from Darlac and Quang Duc offers, "Too often the priorities of RVN officials and American counterparts are at a greater or lesser degree of variance, a factor which mitigates against optimum coordination of effort." The account provided from Quang Ngai suggests the level of tension that existed between the Americans and their Vietnamese counterparts. "I am, as usual, disgusted with VIS lack of co-ordination and ability to work with other Vietnamese agencies as a general rule." The report cites complaints made by the VIS chief about the RVN national police and observes, "but just *try* and get the little guy to do a favor for the police and in this way curry police favor and help! The VIS chief won't budge." During the preceding ten months in Quang Ngai, the report explains, the two previous VIS chiefs stepped down, the last after being caught running off with three hundred pounds of cement that belonged to the government.[25]

TOOTHBRUSHES AND TELEVISION SETS

To divert Vietnamese from such incidents of RVN chicanery, the JUSPAO put in motion an old-fashioned bread and circus routine that included a "specially constructed 'showboat'" that operated on canals and rivers. The boat included a stage about seven meters long and two meters wide. It measured large enough overall to carry a drama team, loudspeakers, projectors, and other communications equipment. In one or two night holdovers at stops along the water near villages, the drama teams gave a long performance consisting of "traditional and anti-Viet Cong songs, together with movies and loudspeaker broadcasts of certain programs taped from Radio Saigon broadcasts." Members of local crowds that assembled to witness the show received numbered leaflets. When officials detected lagging interest in the production, they drew a number, rewarding the leaflet holder with the corresponding digit "a small number of Civic Action gifts—soap, a hairbrush, toothbrush or comb, a child's shirt, etc."[26]

While the JUSPAO dispatched its drama teams via all available means into the countryside bearing any number of gifts, the medium it placed the most hope for enhancing the image of the RVN remained television. Beginning in 1965, American officials stood ready "to assist [the] RVN in establishing TV system by most expeditious means possible." They expected the technology to "have most favorable impact possible." In order to do so, they recognized as important that the "Vietnamese channel appears to be controlled and operated by RVN and that political risks be minimized," and that "RVN telecasting have no audible or visual indication it is sponsored by US facilities." USAID procured an initial thirty-five hundred television sets for distribution and also took on the responsibility of maintaining their operation. The JUSPAO assumed control over the RVN channel's administration.[27]

Within three years, Americans had inundated the Vietnamese countryside with television sets. They distributed well over one hundred thousand units with "approximately 2200 sets in pagodas, public meeting halls and other central locations where large groups can congregate." In addition to the three land-based transmitters in operation by that time in Saigon, Can Tho, and Hue, the Blue Eagle airplane that began service in 1965 continued to provide twenty-three hours per week of pre-recorded programming that reached an estimated seventy percent of Vietnam's population.[28] The JUSPAO boasted of the success of its program. It highlighted incidents such as one that occurred in Saigon in 1966 that resulted in a stabbing that took place as Nguyen Van Hai "tried to fight his way nearer the TV set but was stopped by an agent who advised him to respect women and children." JUSPAO director Barry Zorthian exploited what occurred as evidence of how passionately the Vietnamese embraced the televisions Americans offered them. He chronicled the incident in a dispatch to the USIA, citing it as testimony "that a great many people are turning out to watch the public viewing sets distributed by JUSPAO in Saigon and nearby communities."[29]

The bulk of the programming that went on the air consisted of "educational" programming that used a format "ranging from soap operas for personal hygiene to musicals with an educational theme."[30] The shows sought to dramatize the political message that the United States sought to impart on Vietnamese and paralleled closely the programming offered by the radio operations of the USIA, the Voice of America (VOA). "Uncommon People" highlighted "individual valor." Examples included "the story of a school teacher working under difficult conditions" and the manager of a "program designed to encourage defection from the Viet Cong." Another feature sought to cast dispersion made by claims of the DRV by covering "the Hanoi region and its problems." Others sought to give Vietnamese a sense of their history, by offering them lessons through the recreation of "legendary Vietnamese heroes and their fight against the Chinese aggressors."[31]

Based on what they viewed as the program's success, American officials pursued an agreement between the RVN and the Japanese government to establish a program for educational television for Vietnamese schools. President Johnson designated USIA director Leonard Marks as the official responsible for brokering a deal. Marks received instructions to act outside his role as USIA director "for obvious reasons: the effort should not be construed as a propaganda mechanism." Negotiations with Japan proved successful. The Japanese insisted that the "project surface as a bilateral Japanese-RVN project" to maximize acceptance in Japan and to "minimize the opportunity for Japanese Leftists to criticize increasing Japanese involvement in an American problem." The Sony corporation began training "25 or so" non-Japanese Asians in Tokyo to serve as advisors. Plans called for the program to commence in classrooms throughout the Saigon and Hue area in the summer of 1968.[32] Due to what it deemed as the "unsettled conditions" in the principal cities following the Tet offensive, however, the Japanese ultimately "deferred their contribution" to the program.[33]

Besides the difficulties Americans ran into in trying to persuade Japan to fulfill its commitment regarding educational TV programming, evidence suggests that, despite the crowds who viewed the material that did air, television did not turn out to be the panacea for PSYOP that officials anticipated. Some Vietnamese intellectual and cultural leaders reported, "Vietnamese Television is extremely bad, and that it has become the basis for a great deal of anti-Americanism." In addition, RVN chief of Vietnamese television, Major Do Viet, brought to the attention of American officials a large selection of newspaper clippings "which ridiculed, criticised and mocked" the Vietnamese channel. Reportedly, "tears came into his eyes" as he asked "what he could do to stop this onslaught."

Much of the criticism leveled against the television programming involved charges of attempting to Americanize the Vietnamese and "creating bad taste in the children." It accused the channel of providing nothing but "pessimistic, diversionary entertainment" that "rarely deals at all with RD [Revolutionary Development] cadres, popular forces, social problems, sacrifices by the people and the army, or other 'realistic aspects' of the situation. It suggested the station did little more than provide support for "the claims of the VC that South Vietnamese media are dominated by Americans." Critics maintained that many people did watch the programming, as the JUSPAO cited in the station's defense, but "only because it is a novelty, it is free and there is nothing else to do." They suggested, in fact, that it is "seen as a corrosive and therefore evil force in society."[34]

Critics also leveled the charge against television programming, as they did against other aspects of JUSPAO effort, that concerns with technique and delivery stood as more important to Americans than the content of the message. They charged USIA officials as "obsessed with the technical aspects" and cited as a result:

> The hardware is new and in place; electronically the station functions; English-speaking counterparts are being trained and therefore, to most American eyes everything is well. In fact very little thought or knowledge has gone into the most important area of all: programming.[35]

As the chief consequence, "television in Viet Nam is contributing toward the breakdown of relations between the United States and the Vietnamese intellectual class."[36]

The JUSPAO retorted that to classify television programming in Vietnam as an evil force in society constituted "simply an expression of immature judgment." The rebuttal it issued to address each of the charges made against its operation, however, hardly amounted to an unequivocal denial of the problems identified. Ron Hull, JUSPAO program advisor, conceded several of the points made by the station's critics. He labeled it as true that "Vietnamese TV leaves much to be desired," and admitted, "Until recently, all programming was in the hands of a young Vietnamese 'cowboy' whose taste was polluted and who demanded kickbacks on performing fees."

Hull also acknowledged that the programming "is distasteful to educated Vietnamese," but he suggested the channel intended from the beginning "to reach and bind the greatest number of Vietnamese, most of whom are peasants." Hull labeled the charge that USIA remained obsessed with technique to the exclusion of substance as "significant." He asserted, though "Until the basic facilities are available in a technically-backward country, we must be preoccupied with the technical aspects."[37]

SMOOTHING RUFFLED FEATHERS

As the JUSPAO struggled to acquire a constituency for the RVN over the airways, it also worked to strengthen United States credibility in the countryside. While doing so, it faced many challenges, a number of which stemmed from American behavior that remained inconsistent with any strategies devised to win hearts and minds. The JUSPAO advisor working in Ba Xuyen and Chuong Thien cautioned, "JUSPAO sees a serious equation taking shape that should be corrected before it becomes a problem, namely, that U.S. presence equals the kiss of death." The official illustrated the severity of the problem by noting that "hamlet dwellers along the Bassac River are growing reluctant to go to market, now that US navy boats Patrol the area."[38]

A report filed from the JUSPAO advisor responsible for Darlac and Quang Duc warned that the local people "appear to be growing more sensitive to American presence, feeling that the role of Vietnamese in Vietnamese affairs is diminishing as the American build-up increases." The people there also considered "that Americans are insufficiently effective in curbing inflation and in influencing the RVN to accelerate reforms."[39]

From just outside of Quang Ngai City, the JUSPAO reported a heavy naval gunfire barrage used by the United States that caused extensive damage to several area hamlets. In chronicling the incident, agency personnel likened American behavior to that of "using a cannon to kill a gnat." They urgently advocated that United States military commanders receive some general guidelines that cautioned them to "at least consider the psychological implications—in Vietnam and around the world—of such massive destruction."[40]

The ramifications of using such heavy-handed tactics are evident in a report issued from Ba Xuyen that identifies one hamlet that came to be "victimized by mistake" in June 1967. Even under increased insurgent operations, the report notes, "Renewed VC activity does not seem to concern the population as much as the possibility of accidental damage by US support elements (especially air and armed helicopters)."[41]

Too often. American officials, even those individuals supposedly concerned with PSYOP, made use of methods presumably because of the expediency they appeared to offer them. Often, they considered the potential impact of their actions on the Vietnamese climate of opinion in a haphazard manner. They incorporated an approach that relied on their ability to deal with the ramifications of their actions either as they went along, or even after they had already acted.

American advisors headed for the field received training that emphasized "Don't tell people to move and not use troops to move them." Instructors further advised, "You have to *force* the move and protect the people while moving." To deal with the obvious fallout from a group of people just forcibly moved from their homes, Americans devised the solution to "give them 'sorry about that leaflet'—there are other leaflets as more approach." They believed winning over the Province Chief always remained the key to resolution. Local official cooperation could prevent "stay behinds" from getting mad at the government if they got hurt.[42]

It is clear that this "ruffled feathers" approach, as Americans deemed it, not only proved inadequate to the task, but was counterproductive.[43] The method also alienated Vietnamese who perceived America as first willing to wreak destruction then expecting to mollify the people whose lives its actions destroyed with token and belated shows of good will. A report from Kien-Tuong identifies that the "most serious demonstration of anti-American feeling" took place as a result of bombing an area previously declared as a "free bomb zone." In total, the attack killed forty-two civilians and wounded thirty-five more. The report observes:

> "Dirt was thrown at Americans who visited the survivors' hamlet on the fringes of tiny Tuyen Nhon District town, there being no stones available" . . . Survivors treated by VC medics shortly after the bombing refused treatment from American medics.[44]

It required little for the strained relations that developed between Americans and people who they purportedly sought to make their allies from erupting into full-scale violence. By the summer of 1967, an estimated four thousand Americans, military and civilian, had descended on the provincial capital of Can Tho. The town in the Mekong Delta region had come to serve MACV as the site of its IV Corps headquarters. Before long, the "problems which inevitably accompany the introduction of large numbers of American troops into small Vietnamese towns" began to surface. A "series of brawls" broke out between Americans and local cyclo drivers, who claimed United States personnel refused to pay high enough fares.[45] The strategy Americans adopted to avoid recurrences of the kind of hostility that plagued Can Tho contradicted the tactics for building friendship and trust offered by the cards they carried in their pockets.

United States officials improved relations between Americans and Vietnamese by adopting policies of strict segregation. They solved the cyclo crisis by declaring that type of transportation off limits to all United States personnel. In Can Tho and elsewhere, they strove to minimize the social interaction of Americans and Vietnamese to the greatest degree possible. Officials encouraged Americans to limit their social excursions in Can Tho to Ben Xe Moi, a two-block area that contained eighteen bars and almost as many houses of prostitution. While officials conceded the area did not represent a "savory one," it possessed the virtues of "relatively small size and lack of prominence." They enforced a strict curfew of ten o'clock, and they worked to prevent any intermingling that might occur between Americans and their Vietnamese allies who served in the ARVN. Typically the area's establishments, all identified by signs posted in English rather than in Vietnamese, could not be accessed by ARVN personnel. When ARVN personnel did receive permission to visit Ben Xe Moi, "Americans were cleared out early to avoid involvement" with them.[46]

At Dong Tam, a city in Dinh Tuong Province about seven kilometers due west of My Tho, MACV established a base covering six hundred acres and housing five thousand American troops. Officials charged with establishing the operation reported in 1967 that "the process of bringing American troops to the My Tho area has been accomplished satisfactorily, and there is almost universal agreement on this." They expressed no hesitancy in identifying the reason that enabled this area to undergo such a smooth transition in becoming the forward staging area for United States military operations in the Mekong Delta. Things went so well, officials believed, because of the virtual wall they had constructed between Dom Tam and My Tho. "The key to bringing about this happy state of affairs was the decision to isolate the Dong Tam base from My Tho. The city is off limits for all—repeat all—personnel of the base." Officials cut off any possibility for tension, or for the development of any fruitful relationships, by completely isolating American personnel from the Vietnamese population.[47]

Americans did little to familiarize themselves with contemporary Vietnam before they arrived, and they did as much as possible to prevent their personnel from gaining an understanding of the ongoing concerns of its people once they arrived there. In doing so, they failed to recognize that the psychological contest they so wholeheartedly committed themselves to winning comprised more than issuing platitudes that possessed merit in the American mind. As they hastily quarantined Americans from the people they ostensibly sought to win over, the notion that they needed to gain the trust of the people whose countryside they occupied continued to elude them. They believed they could build the Vietnam they knew would be best for the Vietnamese provided they could prevent the Vietnamese from getting in their way.

While the policy of segregation officials adopted generally succeeded in preventing outright hostilities from occurring between the two allegedly allied groups, it did little to increase the success of American PSYOP. As indicated by one PSYOP Officer with the 5th Special Forces Group (Airborne), 1st Special Forces, Nha Trang, its practitioners in the field believed that "face to face operations on the ground are the most effective form of psychological operations, and that aerial distribution of leaflets is the least effective and aerial loudspeaker operations the next least effective."[48] This strategy hardly proved compatible with the American Way of War in Vietnam, which utilized technology that minimized actual contact between Vietnamese and Americans. In his debriefing that followed his tenure as United States Army Commanding General, XXIV Corps (June 18, 1970—June 9, 1971), Lieutenant General James W. Sutherland Jr. reflected that "while the helicopter" proved to be "invaluable," there existed a "growing tendency to place too much reliance on it." Sutherland observed:

> There is a great inclination to move forces from one point to another, not searching any one area thoroughly and overflying both the enemy and the civilian populace, both of whom would be affected by our physical presence on the ground.[49]

In an interview he gave after the war, JUSPAO director Barry Zorthian recalled a meeting he once had with a South Vietnamese general that took place on an American nuclear-powered aircraft carrier. He pointed out to his interviewer, "You know, an aircraft carrier in full operation is a very impressive piece of equipment," and he underscored the impact America's technological marvel made on his Vietnamese colleague. "This poor Vietnamese," he clarified, "was my counterpart, he was a general, he was minister of information there—just shook his head and said in French, 'Formidable, formidable,' as he looked at this thing."

Certainly, Zorthian's reaction should not have been unexpected from anyone who confronted the awesome display of American firepower at close range. From Zorthian's perspective, though, the bewildered Vietnamese simply did not measure up to the task. He concluded:

> Well, we were "formidable" all over the place, we had so goddamn much gear, we took that Cam Ranh Bay, which was a sand spit, nothing there, sandy beaches and desolate country and so on, and built a major port out of it in a very short time, things like that.[50]

While, in retrospect, Zorthian wondered "whether we could handle that kind of war is a question I really have," he also implied Americans might have found more success had the South Vietnamese simply stepped aside. "I used to say," he remembered, "there was a Vietnamese way of doing things and there's an American way of doing things."[51] In the minds of Zorthian and other American officials, never would the two manage to meet on common ground.

THE OTHER WAR

Regardless of what they said, Americans acted as if they considered their pursuit of the military war, and their conduct in what they deemed as the "other war" for Vietnamese approval, as unrelated. The fighting took precedence in Vietnam, where American officials found it easier to tally the number of communists killed than count how many hearts and minds they won. The former provided them with more concrete evidence than the latter of progress. PSYOP came in a distant second to military endeavors and tellingly fell under the category of pacification, hardly a term that implies a satisfied constituency.

A pacified population is not one that is converted in its allegiance. It is a people made quiet. Officials did not grasp the contradictions their actions afforded their words. As they promoted democracy, they forcibly removed RVN opposition. They carried out what they called a "pacification offensive," which aimed to "attack the VC infrastructure" and to "expand and improve territorial security."[52] Officials considered a pacified group to occupy an area made "secure" through the use of force. They showed little concern with what resided in the hearts and minds of the people living in the areas that they secured.

In Vietnam, Americans pursued strategies and promoted a government that should have struck them as counterproductive to the democratic and free Vietnam they purportedly espoused. One JUSPAO representative who recognized the difficulties inherent to winning the hearts and minds of a pacified population, as opposed to one that felt engaged, observed, "Corps says 'pacification.' Division says 'kill-count.' Conflicting missions necessarily limit all achievements, building and killing as well."[53]

A contest of priorities might have been apparent to some PSYOP personnel, but it is evident that other Americans emphasized eliminating insurgents over winning them over. The JUSPAO routinely faced staffing shortages, budget cuts, and being left in the dark about American policy. Director Barry Zorthian complained to American Ambassador William J. Porter of being omitted from "distribution for messages" and of being denied "access to important substantive messages." He expressed his inability to "function waiting in anterooms, begging members of the staff for bootleg reading of important cables" and grumbled he did not "want to spend half my energies battling for the means to do my job."[54]

The pervasive lack of respect meant budgets often shortchanged the JUSPAO, hampering its operations. In 1965, USIA director Carl Rowan called American support of PSYOP in Vietnam "penny ante in comparison with our expenditures in other fields." He pointed out the USIA program "barely exceeds the cost of one day's operations in the military and economic fields," no matter that he had doubled the program in the past ten months.[55]

Little changed for the USIA over the course of America's time in Vietnam, despite the regular protests made by the agency. Two years after Rowan's remarks, one advisor criticized a freeze on JUSPAO funds as "ill conceived." He pointed out nothing could substitute for the continuing programs his province "must now curtail, and the rapid exploitation of propaganda opportunities which now must be delayed." He labeled the lack of support the JUSPAO received as "disheartening," and he suggested the most recent budgetary cuts to hit PSYOP left local programs "weaker now than they were in 1965."[56]

The JUSPAO constantly found itself in the position of scrambling for resources from both the American and the Vietnamese sides. In May 1967, Zorthian reported to the Mission Council that "psychological operations have run out of money." He noted, "The previously used sources in the imprest funds of USAID and in RVN trust funds are no longer available for psyops use." He also suggested the RVN increasingly had prevented allocated money from reaching the province level, using it for "Saigon programs with a political orientation as opposed to programs to be conducted in the field."[57]

A study conducted of the Chieu Hoi program, supposedly the centerpiece of the PSYOP program, reveals the initiative spurred little more enthusiasm among Americans than it had the RVN. It alleged, "Despite its firm endorsement by top American policymakers, the program has suffered from lack of concrete evidence that the United States is as seriously committed to it as to other efforts of the government." It commented that American support for Chieu Hoi "tended to be limited to supplying materials for the building of reception centers."[58]

JUSPAO officials often lamented, "A common attitude has grown up in Vietnam, primarily among Vietnamese but observable among Free World forces also, that the best Viet Cong is a dead one."[59] American PSYOP personnel not only faced the task of persuading the enemy. They also discovered it necessary to convince individuals on their own side of the value contained in their efforts and in the "other war" overall. Foremost among their skeptics stood America's military. Robert W. Komer, President Johnson's special assistant on non-military activities in Vietnam, identified "the recurring complaint that JUSPAO can't get planes to drop Chieu Hoi leaflets because of tactical priorities." He queried:

> Why don't we ask MACV and ARVN to give the same top priority to Chieu Hoi support they give to tactical operations? In effect, Chieu Hoi demands should always be met. It is just so much cheaper to defect VC than to kill them that this is only commons sense.[60]

This situation did not bode well for PSYOP during the process of "Vietnamization," which President Richard Milhous Nixon articulated as the new American policy in Vietnam beginning in July 1969. In essence, this plan called for Vietnamese to assume all ground operations with material, training, and technology provided by the United States. As historian Marilyn B. Young observes, Nixon's strategy amounted to "a mercenary army on the ground combined with American airpower."[61] This new phase of the war diminished the importance of hearts and minds even further as Americans greatly reduced their support for PSYOP while their Vietnamese counterparts demonstrated little interest in picking up the slack.

PSYOP advisor C. D. Brown reported in December 1970 from the Rung Sat Special Zone, a hotbed of insurgency that spread over 400 square miles between Saigon and the South China Sea, that United States support of operations "has decreased considerably as an effort to speed up Vietnamization." He noted, "U.S. advisors can no longer order leaflets." As a result, it had become "almost impossible to get quick reaction type leaflets." Moreover, Brown observed that it had been "two and one half months since we've last had safe conduct passes or a fresh supply of General Chieu Hoi leaflets." Brown continued to call it "most unfortunate" that "air assets have been cut back and equally so that the Commanding Officer of the Rung Sat doesn't think psyops important enough to warrant use of VN water-borne assets."

Brown remained optimistic, offering that, "if given a chance, the system should work." He worried, though, that there remained "a reluctance on the part of the PSYOP/Pol War unit leaders in exercising this system and insist on coming to the U.S. advisor as in the past." This situation did not result from "ignorance," Brown insisted, because "for each instruction I've received that has told me what I can't do any longer, they've received one telling them of their new inherited responsibilities and how to execute them."[62]

In addition to the difficulties the agency faced convincing others of the merits of its programs, throughout its existence, the JUSPAO had to contend with profoundly divergent views about PSYOP espoused by its military and civilian personnel. In an interview he gave after the war, United States Army General Donald D. Blackburn cites the persistent "difference in philosophy on PSYOPS between the military and the civilian side of the government" as undercutting American efforts in Vietnam. Blackburn, who laid the groundwork for the JUSPAO in cooperation with USIA director Carl Rowan, observed "you could see the weakness in it before we even started."

While civilians in the JUSPAO pleaded otherwise, Blackburn claims PSYOP to them implied a "distortion of the truth." Military personnel reflexively objected because they believed their product had "to be truthful." True to the use of PSYOP within the American Way of War, commanders such as Blackburn had little use for the political brand of PSYOP civilian officials preferred. For them, the American style of PSYOP served to save American lives by securing enemy surrender. In order for it to accomplish that task, to work effectively as a tactical force multiplier, it stood as imperative that American PSYOP maintain a high level of credibility. According to Blackburn, due to their philosophical differences, military and civilian personnel proved incapable of ever fully "meshing" into one coherent effort.[63]

Meanwhile, JUSPAO civilians considered their military counterparts amateurish. Zorthian contends during the war that America's civilian propagandists measured their skill against "real communications problems in a real world." Military officials, however, remained "submerged to a junior level to plan, train, and farm out motivational research." He accuses the military of spending its time "ruminating over critiques of past wars" instead of operating in reality. Zorthian identifies what he considers as the "crux of the problem" when he cites the necessity of clearly demonstrating "psyops is an asset not a detriment to career officers." In Vietnam, military officers assigned to psyops functions came and went in "unending succession." Zorthian points out, "Without an in-service identity, recognition and status, there was and is little incentive for the competent officer to remain in Army psyops assignments." Blackburn and Zorthian's comments make it abundantly clear that one of the JUSPAO's greatest obstacles stemmed from the deep "sub-surface frictions" existent among its own ranks.[64]

KILL THEM ALL

Throughout its history, the United States military has distinctly preferred destroying its enemies to either converting them or persuading them to surrender. When faced with carrying out PSYOP in Korea, it responded by pursuing these activities in a strictly subordinate way. This trend not only continued in Vietnam, it intensified as Americans fighting and advising in the field proved increasingly unwilling to discern between friend and foe and helped perpetuate an animosity that gave credence to cries of imperialism made by the RVN's opposition.[65] The American military went into Southeast Asia unaccustomed to thinking about psychology as anything but a sporadically useful prop for its conventional operations.

Confident in its ability to overwhelm opponents with superior force, the United States military saw little reason to understand the motivations of the enemy it faced. It possessed a formal conception of warfare, one that involved organized troops, front lines, and uniformed soldiers. America's military failed to abandon its way of thinking in Vietnam, even when confronted by the realities of the insurgency it faced. It went there with the impression insurgencies constituted a military version of amateur hour and refused to concede its vision.[66] It recognized the resourcefulness and the organization of the guerrillas it fought against, but it refused to consider that their strength derived from the ideas that motivated both them and the Vietnamese who came to their aid.

The artificial constructs America's military had created to separate psychological from conventional operations prevented it from appreciating the interconnectedness of ideas and actions in Vietnam. It made no attempt to understand the forces at work in Vietnam and to devise a strategy that might rob them of their power to promote discontent with American and RVN actions. As reflected in the evidence collected by Eric Bergerud, the psychological implications of their actions proved the last thing on the minds of typical American soldiers in the Vietnamese countryside.

Jay Lazarin, a rifleman in the 25th Infantry Division of the United States Army in 1967, recalls, "The friendly civilians were seen as no such thing. To us, there were no friendly civilians, only ones who posed no immediate threat." According to Thomas Giltner, who in 1966 served as a platoon leader in the 25th, "It was the hallmark of the war: You could never tell who was the enemy and who was not. Therefore, you treated everybody with suspicion and distrust."[67] These soldiers succinctly stated in their own words the official view of both military and government officials. The only good Vietnamese was a harmless one, one that had been pacified.

Despite the cards they carried in their pockets reminding them to respect Vietnamese civilians, some American soldiers acted on their frustrations regardless of the potential psychological ramifications. A saying made popular among American forces by the Marines advised, "Get the people by the balls, and their hearts and minds will follow."[68] The JUSPAO advisor in Quang Nam reported that the 39th Rangers apparently operated routinely according to that maxim. He recounted that the outfit's members "again displayed their inability to conduct themselves properly on operations." His report noted that the troops "continually took food from the populace without any payment" and that a village chief reported "the battalion threatened to assault a woman."[69] Not withstanding the number of leaflets dropped that explained the virtuous motives behind the American presence, even remote instances of such conduct of American troops invited successful exploitation by the insurgents.

Whatever the average soldier felt about Vietnamese psychology and despite the military's tradition of apathy regarding PSYOP, a movement had begun by the end of 1966 to wrest control of the program in Vietnam from the civilians who ran it. Communist success in Vietnam had been widely attributed by Americans from the beginning to be the product of organization.

After operating for over a year through the JUSPAO, and for over a decade overall, the USIA had not made sufficient enough progress in eroding the insurgency's substructure.

For impatient officials in Washington, winning hearts and minds had become increasingly about pacification. To achieve that condition, they had grown convinced of the essential prerequisite of security. By the end of 1966 for certain, it became evident to the Johnson administration that this job should be assumed by the military. The USIA should concentrate all of its energy on creating a more favorable impression of the war at home and abroad.

NOTES

1. Report, TAY NINH (III CORPS), September 20, 1966, "Revolutionary Development," 1966–68, Lee Papers, MHI.
2. Report, DARLAC and QUANG DUC Provinces.
3. Report, "II CORPS NHATRANG," ibid.
4. Memo, for the President, May 1, 1967, "Vietnam Memos (A) Volume 70 April 25–May 8, 1967," box 43, National Security File (NSF)—Country File—Vietnam, LBJL (emphasis in original).
5. Report, II FFORCEV/BIEN HOA, September 20, 1966.
6. Report, Chieu Hoi Information and Guidelines, June 1966, "Chieu Hoi (3, 2 of 2)," box 2, NSF—Komer—Leonhart, LBJL.
7. Memo, "A Report on South Vietnam," from the Director of the Bureau of Intelligence and Research (Roger Hilsman) and Michael V. Forrestal of the National Security Council Staff to the President (John F. Kennedy), January 25, 1962, *FRUS, Vietnam: January–August 1963*, III: 55.
8. Newsletter, Assistant Director for Chieu Hoi to all Regional Directors and Provincial Representatives, Undated, "Vietnam Correspondence," box 3, NSF—Files of Robert W. Komer, LBJL.
9. Report, Improving Effectiveness of Chieu Hoi, September 1967, "Vietnam: Simulatics Corporations: Improving Chieu Hoi," box 239, NSF—Country File—Vietnam, LBJL.
10. Report, BA XUYEN/CHUONG THIEN Provinces, September 20, 1966.
11. Report, HAU NGHIA PROVINCE, September 20, 1966.
12. Report, BINH-TUY/LONG KHANH Provinces, September 20, 1966.
13. Letter, Ogden Williams to Richard Holbrooke, August 13, 1966, "Chieu Hoi (3, 2 of 2)," box 2, NSF—Komer—Leonhart File, LBJL.
14. Memo, Carl Rowan to the President (Johnson), March 16, 1965, "Rowan Report," box 190, NSF—Country File—Vietnam, LBJL.
15. Report, TUYEN DUC/LAM DONG PROVINCES, September 20, 1966.
16. Memo, Memorandum for the Record, Meeting with Colonel Anh, August 10, 1966, "Chieu Hoi (4)," box 3, NSF File, Komer—Leonhart File, LBJL.

17. Memo, The Status of Non-Military Actions in Vietnam, November 20, 1965, 41 Point Program in Non-military sphere, box 198, NSF File—Country File—Vietnam, LBJL.

18. Letter, Ogden Williams to Richard Komer, January 6, 1967, "Vietnam Correspondence," box 3, NSF—Files of Robert W. Komer, LBJL.

19. Letter, to Barry Zorthian from Robert Komer, June 1, 1966, "Psychological Warfare," box 21, NSF—Komer—Leonhart File, LBJL.

20. Transcript, Barry Zorthian interviewed by Ted Gittinger, May 26, 1982, 84–64: 10, Oral History Collection, LBJL.

21. Letter, to Barry Zorthian from Robert Komer, June 1, 1966.

22. Letter, Ogden Williams to Richard Holbrooke, August 13, 1966.

23. Report, HAU NGHIA Province, September 20, 1966.

24. Report, HAU NGHIA Province, September 20, 1966.

25. Report, QUANG NGAI/QUANG TIN Provinces (emphasis in original).

26. Article, "USIS-VIS 'Showboat' Takes Message to Vietnamese Villagers," *USIA Correspondent*, November 1965, "USIA Correspondent 1965–1967," box 14, Papers of Leonard Marks, LBJL.

27. Telegram, to AmEmbassy Saigon Joint State/DOD/USIA/AID Message, Television for Vietnam, November 17, 1965, "Educational TV," box 30, LBJL.

28. Memo, Biweekly report for the President, July 3, 1968, "USIA—1968," box 135, Confidential File—Agency Reports—USIA, LBJL.

29. Field Message, JUSPAO to USIA Washington, Television Claims its First Victim in Viet-Nam, February 23, 1966, "JUSPAO," box 30, Papers of Leonard Marks, LBJL.

30. "Need Nielsen to Count Vietcong Eavesdroppers on Ky's ETV Weapon," *Variety*, December 14, 1966, 1.

31. Article, "Voices of Vietnam," *USIA Correspondent*, March 1966, "USIA Correspondent 1965–1967," box 14, Papers of Leonard Marks, LBJL.

32. Report, Educational TV in Viet-Nam, undated, "Educational TV," box 30, Papers of Leonard Marks, LBJL.

33. Memo, Biweekly Report for the President, July 3, 1968.

34. Memo, Television in Viet Nam, April 25, 1967, "Educational TV," box 30, Papers of Leonard Marks, LBJL.

35. Memo, Television in Viet Nam, April 25, 1967.

36. Memo, Television in Viet Nam, April 25, 1967.

37. Memo, Television in Vietnam, June 8, 1967, "Educational TV," box 30, Papers of Leonard Marks, LBJL.

38. Report, BA XUYEN/CHUONG THIEN Provinces, September 20, 1966.

39. Report, DARLAC/QUANG DUC Provinces, September 20, 1966.

40. Report, QUANG NGAI/QUANG TIN Provinces, September 20, 1966.

41. Report, BA XUYEN/CHUONG THIEN Provinces, September 20, 1966.

42. Hand written notes, Relation with Counterpart, May 12, 1966 (emphasis in original).

43. Report, PHUOC TUY/VUNG TAU Provinces, September 20, 1966.

44. Report, IV CORPS (CANTHO), September 20, 1966.

45. Cyclos are pedicabs common in Vietnamese cities.

46. Memo, American Presence in Can Tho—The Ben Xe Moi'Area, December 8, 1967, "Vietnam Special Joint Reports—IV Corps," box 232, NSF—Country File Vietnam, LBJL.

47. Memo, American Presence in My Tho, Dinh Tuong Province, December 7, 1967, "Vietnam Special Joint Reports—IV Corps," box 232, NSF—Country File Vietnam, LBJL.

48. Report on Psychological Operations Intelligence in Vietnam, November 17, 1967, Folder 01, Box 14, Douglas Pike Collection: Unit 03—Insurgency Warfare, VA-TTU.

49. Debriefing of Lt. Gen. James Sutherland, June 1, 1971, Folder 06, Box 17, Douglas Pike Collection: Unit 02—Military Operations, VA-TTU.

50. Transcript, Barry Zorthian interviewed by Ted Gittinger, May 26, 1982, 84–64: 12–13, Oral History Collection, LBJL.

51. Transcript, Barry Zorthian interviewed by Ted Gittinger, May 26, 1982.

52. Report, Pacification in South Vietnam During August 1967, October 1967, "Revolutionary Development Program 1 (March–October 1967)," box 59, NSF—Country File Vietnam, LBJL.

53. Report, Special Joint Report on Revolutionary Development, July 31, 1967, "Vietnam, Corps IV, Kien Hoa," box 232, NSF—Country File Vietnam, LBJL.

54. Memo, Barry Zorthian to the Ambassador (Lodge), April 8, 1966, "Material on JUSPAO," box 4A39, Papers of Barry Zorthian, LBJL.

55. Memo, Carl Rowan to the President (Johnson), March 16, 1965, "Rowan Report," box 190, NSF—Country File Vietnam, LBJL.

56. Report, Special Joint Report on Revolutionary Development, July 31, 1967, "Vietnam, Corps IV," Kien Hoa, box 232, NSF—Country File Vietnam, LBJL.

57. Memo, Mission Council Action Memorandum No. 196, May 26, 1967, box 41, "NSF—Country File Vietnam," LBJL.

58. Report, Observations on the Chieu Hoi Program, January 1966, "Chieu Hoi (3) [1 of 2]," box 2, NSF—Komer—Leonhart File, LBJL.

59. Newsletter, Assistant Director for Chieu Hoi to all Regional Directors and Provincial Representatives, Undated, "Vietnam Correspondence," box 3, NSF—Files of Robert W. Komer, LBJL.

60. Memo, Memo for Assistant Secretary of Defense for International Security Affairs John McNaughton, February 6, 1967, "Psywar," box 8, NSF—Files of Robert W. Komer, LBJL.

61. Young, *The Vietnam Wars*, 240.

62. Memo, from Psyops Advisor Rung Sat Special Zone to Senior Advisor Rung Sat Special Zone, December 16, 1970, folder 02, box 01, Carl Nelson Collection, VA-TTU.

63. Transcript, United States General Donald D. Blackburn interviewed by Lieutenant Colonel R. B. Smith, 1983, 335–38, Senior Officers Oral History Program, MHI.

64. Letter, Barry Zorthian to Murray Dyer, November 13, 1966, "Material on JUSPAO," Papers of Barry Zorthian, box 4A39, LBJL.

65. Cincinnatus Currey, *Self-Destruction: The Disintegration and Decay of the United States Army During the Vietnam Era* (New York: W.W. Norton, 1981).

66. Cincinnatus Currey, "Preparing for the Past," *Military Review* 69:1 (January 1989): 3–13; Douglas Blaufarb, *The Counterinsurgency Era* (New York: The Free Press, 1977); Andrew F. Krepinevitch Jr., *The Army and Vietnam* (Baltimore: The Johns Hopkins University Press, 1986); J. D. Waghelstein, "Post-Vietnam Counterinsurgency Doctrine," *Military Review* 65, no. 5 (May 1985): 36–45.

67. Bergerud, *Red Thunder*, 222–23.

68. *The Pentagon Papers*, Gravel edition, II: 529.

69. Report, Quang Nam (I Corps), Special Reports to Ambassador Porter, August 19, 1966.

Chapter Ten

Maximum Exposure

OVERVIEW

"Courting Support" is the first of this chapter's four sections. It indicates that the importance of building a consensus for American policies in Vietnam quickly became the priority of the JUSPAO under a Johnson administration that anxiously monitored its approval rating. The USIA faced increasing pressure to secure more favorable press coverage by diverting attention away from military battles and toward "the Other War" in Vietnam. The second section, "Pacification: Enhanced and Expanded," illustrates the American abandonment of winning Vietnamese hearts and minds as a means to bring about stability. The new policy marked yet another wave of oppression for Vietnamese hearts and minds that Americans had deemed as tertiary to the conflict long ago but now literally moved to set aside.

This chapter's third section, "A Czar Undressed," examines the dismantling of the JUSPAO by the USIA. The USIA approved of turning PSYOP over to the military, not because it objected to the blurring of information and propaganda that it performed in Vietnam, but because it felt its mission to advocate America's policies in Vietnam to other people of the world had been compromised by a lack of resources. The chapter's final section, "What We Learned," contends that the USIA had become little more than a tool that the Johnson administration proved willing to use for its own political benefit. While the USIA turned psychological warfare over to the military in Vietnam and pledged to stay clear of such activities in the future, its continued interest in manipulating the press indicated the agency still conflated its selling of impartiality with the journalistic ideal of objectivity.

COURTING SUPPORT

By the beginning of 1966, the United States had committed nearly two hundred thousand troops to Vietnam, and it had waged a ten-month bombing campaign against the DRV. With no end to the conflict in sight, American officials were anxious to demonstrate that their nation's effort had made some sort of positive impact. They had recognized early that not only did the United States need to win Vietnamese hearts and minds, but that it also had audiences to court in other nations and at home. In October 1965, the United States Mission in Saigon received a note from Washington encouraging it to do more than it currently did to promote the conflict's civil dimension. The cable called attention to the "continuing concern at the highest levels" of the "need to emphasize our non-military programs in Vietnam and give them maximum possible public exposure."[1]

The Johnson administration considered the matter so urgent that, by the beginning of the next year, it had hastily convened a conference in Honolulu to discuss how best to promote America's nonmilitary initiatives in Vietnam. The delegation of United States and RVN leaders in Honolulu ultimately reached a consensus that stressed, "Building democracy in rural areas is as important as military battles." The participants at the meeting also agreed that every effort needed to be taken on both the American and Vietnamese sides to publicize their endeavors beyond the fields of battle. The conference called for Vietnamese leaders to make more American television appearances, while President Johnson personally urged both governments to "stress at every possible opportunity work being done in increasing school enrollments, building schools and increasing the number of health teams."[2]

Immediately, the USIA set to work on the directive under the stewardship of its new chief, Leonard Marks. Marks had been chosen by Johnson to replace Carl Rowan, who had resigned late in 1965 over his feeling that he was left out of Johnson's policy-making circle.[3] Marks came to the USIA with a combination of administrative, legal, and communications experience. He had his own radio program in Pittsburgh, Pennsylvania, at the age of 16 and went on to graduate from the University of Pittsburgh's law school. During the Second World War, he became the assistant to the Federal Communications Commission's General Counsel. In 1946, he founded the law firm of Cohn and Marks. Marks had enjoyed a close personal relationship with Johnson for years, prompting speculation in the press that his appointment could be a negative for the nation. The media worried that Marks might utilize the USIA as "merely an instrument of the president's personal will in telling America's story as he wants it told." The agency's new deputy director under Marks, Robert Wood Akers, was also a longtime friend of Johnson. He had known the president for over thirty years, serving as editor-in-chief of the Beaumont, Texas, *Enterprise*.[4]

Marks dutifully reported back to the president following the meeting in Honolulu, "All of our media, including the Voice of America, gave this meeting top priority." The agency had placed all of its "emphasis" worldwide on the "social and economic side of our effort in Vietnam." It noted in particular that special arrangements had been made "to speed delivery of press output in Saigon to the Department of Agriculture." The agency reported making an immediate impact, citing the "record pamphlet distribution of 28,000 copies of 'Why Vietnam'" by USIA/The Hague with one thousand additional requests and "an expressed approval of U.S. policy in Vietnam." The agency also pointed to correspondence it received from a Finnish teenager that "sums up what many Finns have written thusly." The youth wrote, "I have always said that Americans are right in Viet-Nam. Now your programs have helped me to understand better what you are doing there . . ."[5]

In addition to its efforts through other posts, the USIA initiated a program that underscored its preference for manipulating the media rather than advising officials for real change. Through an innovative operation called the "Third Country Journalist's Program in Viet Nam," the USIA began actively recruiting news organizations from around the world that employed journalists it coveted to report on the war. The JUSPAO manifested its willingness to take care of reporters whose organizations proved "unable or unwilling to pay the expenses" for a trip to Vietnam. It did so by "supplying transportation and, in some instances, per diem" to cover their costs. The reporters that participated in the program provided the favorable reports that American officials had hoped their initiative would elicit. They confined their output to reports based on their "first hand-knowledge," gained under JUSPAO supervision, of the "non-military activities of the U.S. Government and efforts being made by the Government of Viet Nam to introduce a democratic society."

During 1966 alone, the USIA provided "transportation or other financial assistance" to 84 journalists. In total, it recruited more than three hundred "third-nation correspondents" that represented one hundred and ninety-two news organizations from forty-two different countries.[6] The impact made by the program on one of these reporters, Clas Johnson, illustrates the success that the JUSPAO achieved by offering incentives to those who would come to Vietnam and hear it present America's case. Johnson served as the Foreign Editor of the *Swedish Press News Service*, and he visited Saigon as part of a group of thirteen European journalists that operated under JUSPAO sponsorship. When his tour of Vietnam with JUSPAO representatives ended, Johnson returned home to pen stories with headlines such as "War-Torn South Viet

Nam Raises Living Standard" and "South Viet Nam's Fight Against Poverty: A Big War That Nobody Talks About."

Desperate for good ink, officials detected no discrepancy between calling for a policy of "maximum candor" out of one side of its mouth and from the other side initiating a program that underwrote the endeavors of reporters who emphasized the American government's view of events. The USIA post in Stockholm boasted about the impact made by the "Journalists in Viet Nam" through Clas Johnson's reporting. "Coming at a time when the pages of the Swedish press were virtually closed to any exposition of the allied position in Southeast Asia, the Johnson series gave the U.S. cause in Sweden its greatest boost in nearly two years." Twenty-three Swedish newspapers ultimately ran Johnson's "strongly pro-U.S." articles on Vietnam.[7]

Another foreign correspondent courted by the JUSPAO, Finnish reporter Lieutenant General Jay Jarventaus, "expressed himself in exuberant terms" on his departure from Vietnam. He stressed to JUSPAO officials his "appreciation for the opportunity to see operations in Viet Nam," and he told them his visit "was highly worthwhile and should result in a series of favorable articles." Jarventaus left Saigon with a stack of photos, maps, and charts under his arm that the JUSPAO eagerly provided him.[8]

The JUSPAO program to seduce foreign journalists to report about Vietnam came under the scrutiny of the United States Senate Foreign Relations Committee in August 1966, when USIA director Leonard Marks received a summons to testify about it on Capitol Hill. Marks satisfied his critics in the Senate by maintaining the program stemmed from good old American generosity. He testified that the USIA paid the bills "because the correspondents' papers could not otherwise have afforded to send a man." Besides, he added, "Foreign newsmen were under no obligation to write anything other than what they saw and felt." Marks contended that his agency's program only sought to "open foreign minds to U.S. problems and policies."[9]

The USIA director felt good about his appearance and commented afterward, "the Committee enthusiastically endorsed the program and urged me to expand it." He informed all area directors that the agency "will continue on the same basis as previously and that there is to be no dimunition of our effort." He even alerted them that, "if circumstances which justify paying transportation of a correspondent for a financially-able newspaper, I will consider such."[10]

Despite the apparent success of these operations, the State Department noted that even the governments of America's closest allies faced growing difficulties in supporting the United States policy in Vietnam. In July 1966, it found "widespread displeasure and disapproval over some of the means used to attain our goals in Vietnam." It declared that there existed "practically no chance" Canada or any Western European country besides France, perhaps, "would come out into open opposition to the U.S." At the same time, though, the State Department study identified "increasing public hostility toward the protraction and escalation of a war often called 'absurd' in Western Europe." The data revealed that "official support" from these states stood as "tepid" at best, "nonexistent" at worst. Updating the findings in March 1967, the department warned, "The erosion of public and official support has been particularly pronounced over the past few months." Hanoi's peace offensive had successfully refocused public attention back to Vietnam's military conflict and away from the "other war" Americans had touted.[11]

PACIFICATION: ENHANCED AND EXPANDED

The movement to remove the JUSPAO from much of its pacification chores and concentrate full-time on public relations got underway by the middle of 1966 and accelerated in 1967. Motivating the Vietnamese, a task Americans both in and out of the USIA viewed as a charade from the onset, would be fully turned over to the military. It took the term pacification literally and viewed the problem as one involving what it did best, providing organization and security.

In October 1966, Robert W. Komer, Johnson's Washington-based pacification guru, served notice. "As I've repeatedly warned privately, we're not yet going anywhere fast." He suggested further, "Nor will we, in my judgment, so long as we fail to recognize that pacification is first and foremost a military task—providing continuous local security." Komer pointed out that it served no purpose to "blame" the civilians when the critical first phase of pacification is primarily a military job. In his view, "Only the US and RVNAF [ARVN]" possessed the resources capable of bringing about a positive turn of events.

Komer urged Johnson: "*Give Pacification Management to Westy* [General Westmoreland]." He continued:

> You've heard my voice on this. I'll just say again that we *won't* get up real momentum in pacifying hamlets until you give it to the only people who can do most of the job. Lack of progress so far is not the fault of the civilians—it is basically because the US and ARVN military haven't provided the one essential ingredient—local security.[12]

Komer believed that his ideas offered the United States "the best chance of a satisfactory outcome in Vietnam." All America needed, he urged Johnson, was "better management, better orchestration, and better use of the assets we already have."[13] Komer emphatically voiced his opinion to other officials over the next six months, so much so that he earned the nickname of the "blowtorch" from Ambassador Lodge for his numerous memos.

Komer possessed no experience with East Asia when Johnson appointed him as the White House's supervisor of America's "other war" in Vietnam in 1966. He began his long government career as a corporal in Army intelligence during the Second World War. After returning home, he took an M.B.A. from the Harvard Business School but then went to work as a senior analyst in the Office of National Estimates at the CIA. He headed staffs on Western Europe, on the Middle East, and on the Soviet Union before joining the National Security Council staff under the Kennedy administration. Komer rose to become the deputy of McGeorge Bundy, President Kennedy's special assistant for national security affairs. In that position, Komer became Washington's specialist on affairs in the Middle East and in Africa. Johnson turned to him for help in Vietnam because of his "brassy, ruthless, work-and-ambition-driven qualities." He was unconcerned about Komer's lack of expertise on Vietnam.[14]

Endorsed by the president, Komer's view carried the day. By the middle of 1967, the United States created the Office of Civil Operations and Revolutionary Development Support (CORDS) under MACV as the single manager for all pacification advisors. Komer received an ambassador's rank and headed the new organization. Throughout 1966, while he had refused to blame them, Komer made clear to the civilians running pacification programs in Saigon his displeasure with where things stood and pushed them to achieve more progress at a faster pace.

In June, Komer wrote to the embassy for Zorthian. "Would appreciate your taking lead in drawing together coordinated set of action proposals to energize lagging Chieu Hoi program." He informed the JUSPAO director, "Highest authorities here have long regarded defector program as major asset." He called on Zorthian for "positive recommendations on ways to reinvigorate program, staffing, and defector rehabilitation."[15]

When the change finally came about, the officials who pushed for it to occur in the first place, particularly Komer, appeared anxious to downplay the increased military element. Shortly before the announcement of the CORDS reorganization took place in Saigon on May 11, 1967, by United States Ambassador Ellsworth Bunker, Komer advised the president on how the new policy should be explained to the public. He warned, "The big criticism to head off is that this means further militarization of the war." He believed that such a view could be "knocked down" by Bunker. He should stress that "pacification is the 'heart of the matter,' and that the key prerequisite to pacification is security."

Komer suggested that Bunker should state that he "does not regard pacification as a civil or military problem, but as a civil/military problem." Further, he believes the change will result in "greater efficiency and economy in this key endeavor." Komer asserted with confidence, "If Bunker will say the above, we'll draw enough of the short-term sting to minimize its impact."[16]

Komer's new creation would push a decidedly more aggressive pacification effort that officials referred to as "enhanced and expanded." The bulk of CORDS' work consisted of developing Ap Doi Moi, a type of hamlet officials believed would "continue to grow stronger politically, socially, and economically as a part of the community of prosperity." Officials stipulated eleven conditions that these hamlets must meet upon their completion. The program lacked new ideas. It recycled the old concept of creating Strategic Hamlets through pacifying the population and eradicating communists. The initiative brought a heightened level of violence, though. Placed under MACV, it militarized the "other war" to bring about Vietnam's evolution into a democratic state. The first standard for an Ap Doi Moi hamlet established that "the underground communists have been annihilated."

Other criteria set down for the program could hardly be construed as responsive to popular needs. The language used to convey the remaining guidelines is militaristic while implying centralized control and tight organization. The tone used makes clear the road to the pacification status American officials envisioned would be paved with violence. The standards called for illiteracy to be "eradicated," hatred "removed," bullies "eliminated," and diseases "attacked." The program's guidelines insisted that the Vietnamese demonstrate their gratitude for those who implemented this agenda. They demanded that American and Vietnamese military personnel be "treated in a deserving manner."

The program called for democratic government and the creation of an "anti-Viet Cong peoples' struggle," suggesting both could result from "organization." It remained vague, never explaining the process for reform beyond the need to organize and eliminate. The initiative also projected a measure of arrogance. Its standards called for the development of a communications network, disregarding the traditional system that had tied villages together for centuries.[17]

A CZAR UNDRESSED

Komer was instrumental in the change that brought about CORDS and the remilitarization of pacification it accompanied. Nevertheless, he hardly stood alone in pushing for the adoption of a new organization and for the use of more direct means. A plethora of studies and reports by late 1966 by the United States Army Staff, the Priorities Task Force, the Roles and Missions Study Group in Saigon, and various other observers and visitors criticized the pacification effort.[18] By the end of the year, a clearly frustrated Zorthian observed that all of the "inspections, reviews, reorganizations and re-programming exercises" that the JUSPAO had endured rendered it "extremely dificult to maintain an orderly organization with any semblance of morale and attention to our program and mission."[19]

One of the more prominent pushes to move in the direction Komer advocated came from the military. The Ad Hoc Committee on Army Psychological Operations expressed its finding that JUSPAO duties fell under Army jurisdiction and pointed to the greater resources the military possessed as evidence it could do a better job than the USIA.[20] Military personnel who served under the JUSPAO also joined in the chorus and emerged as some of the agency's harshest critics. After his assignment to the JUSPAO concluded in the summer of 1966, United States Army Brigadier General John F. Freund commented extensively on his experience.

Freund began his review by identifying the "small, but significant, percentage of bright young men" in the agency's service that possessed "the capacity, self-discipline, and desire to absorb graduate-level training." He then lamented, "The aspirations of these men are largely ignored, and there is almost a premeditated effort" to "take the men of unusual caliber and beat them back into the mold of the average." Freund believed JUSPAO standards of professionalism left much to be desired and the organization suffered commensurably.

Freund contended that, measured by "only what JUSPAO has done," its record stood as "excellent." Yet, when examined by "what JUSPAO has done with relation to the requirements at hand," the agency's progress had been "unsatisfactory." He underscored the existence of "*a pressing requirement to re-examine and to redefine the relationship between JUSPAO and MACV.*" He stressed that the JUSPAO would be better equipped to carry out its duties in the countryside through a realigning with MACV that would eliminate its current fragmentation of command. JUSPAO functioning suffered because of inadequate management, according to Freund, who advised that a single director could not properly oversee both PSYOP and public relations. He called for the day-to-day operations of the agency to be supervised by a director wholly concerned with PSYOP. The post should be filled by the general officer assigned to the JUSPAO.

Most telling of Freund's observations is his assessment of the JUSPAO's reluctance to incorporate Vietnamese perspectives into its campaigns. He called the relationship between American and Vietnamese JUSPAO personnel "an area of major deficiency." He contended that the Advisory Division of the

MACV PolWar Advisory Directorate stood as "physically integrated" with the General Political Warfare Department of the ARVN High Command. the JUSPAO, however, maintained its distance.

To the extent that the JUSPAO had been constructed to provide a "source of information relating to the psychological needs, reactions and characteristics of the people, as well as for the development, distribution, and personal contact measures for responsive actions," it remained sorely lacking. The Field Representatives Division and Development Division of the JUSPAO theoretically held these responsibilities. Freund alleged, nevertheless, "There is not a single space in this office to perform the vital functions of correlation, analysis, research, or short range planning." Worst of all, this situation "was discovered and reported" within months of the agency's creation but had yet to be addressed nearly a year later.[21]

JUSPAO Acting Director Harry Casler, during a brief leave of absence taken by Zorthian, responded to Freund's allegations. He contended that it was not feasible, "given our primary mandate and our limited resources, to compete in terms of output or productive capacity with the vastly larger U.S. Army." He conceded, however, that "tThe problem of JUSPAO-RVN relationships is a real one." He identified the agency's need to "work toward a community of effort and a joint approach to common psyops problems," and he observed a study group had already been formed to tackle the problem.

Casler acted in a way that was consistent with the JUSPAO's past record. He blamed the Vietnamese for the agency's problems. "The Vietnamese," Casler alleged, "relegate it to a low level of coordination; they are reluctant to pass on our materials, and they make no effort to show us their materials." Again, he cited more efforts already under way by the JUSPAO to correct the situation.[22]

Another report Casler filed a day later to report on progress made by the JUSPAO to support pacification further indicates how far removed from its Vietnamese audience, and the RVN, for that matter, the agency remained. Casler indicated his belief that the report, which primarily quantified the items the JUSPAO produced for distribution, such as posters (521,400), leaflets (2,224,000), slogan sheets (30,000), and calendars (600,000), represented "an impressive contribution" to the latest American-initiated program of pacification, Revolutionary Development (RD). Still, he acknowledged the need to worry about "the possible danger of JUSPAO's information support running ahead" of the Vietnamese and of his agency "replacing VIS as the major contributor to information activities." In terms of his audience, he conceded, "Frankly, we do not know what the peasants, even those whose lives have been affected by it, think of the Revolutionary Development program." He did not possess high hopes, though, as he cited the "inherently suspicious nature of the peasant."[23]

The JUSPAO admittedly struggled with its PSYOP chores as it tried to maintain control over them; and its information effort continued to receive criticism. One official from Washington who visited Vietnam in May 1966 acknowledged that the press corps in Saigon was "young, inexperienced, and sensationalist minded." He assessed it as a "difficult group to work with at best," but he remained "convinced that more can be done."[24] In August, the USIA director wrote to Zorthian about his appearance before the United States Foreign Relations Committee. While he found the Committee "courteous and considerate," it also proved "genuinely concerned about the news and information policy."[25] Meanwhile, observers of the situation in Saigon began to identify Zorthian and the two hats he wore on his head, public relations and PSYOP, as the fundamental problem.

One review of the JUSPAO called Zorthian "undoubtedly one of the Agency's best officers" but noted the "stresses and strains" of his situation. The study suggested that Zorthian "may have lost some of his objectivity" in the difficult circumstances where he worked. It would likely be "to the Agency's advantage," the study concluded, "if he were reassigned as soon as is practicable."[26]

A few months later, White House staff member Richard Holbrooke provided an even more forthright assessment of Zorthian's position to Komer. On the eve of creating CORDS, Holbrooke advised that whatever else happened in the coming months, the "JUSPAO should not be allowed to survive much longer." He reasoned there stood "no logical reason for press and psywar to be combined in a single organization, any more than press and health, or press and education." A "more rational breakout of the duties now assigned to JUSPAO," in his judgment, "would be to split *Press* and *Psywar* completely."

Holbrooke attributed the JUSPAO's creation to the war's timing and to "the incredible bureaucratic wizardry of Barry the Z." Nevertheless, he continued, while "Zorthian swiftly built up his empire," the JUSPAO director spent his time "mostly with the American press, and he has never been able to give psywar the attention it deserves." Instead, Zorthian "stymied staff initiatives, has not made a real effort to work closely with the RVN, and has not built up a quick reaction outfit." Regardless of what the future held, Holbrooke implored Komer to "abolish the single Minister for Information, sitting above both press and psyops below the Ambassadors, interfering in everything but not doing his job (at least his psyops job) the way it should be done."[27]

WHAT WE LEARNED

Under pressure from all quarters for change, and never sold on the possibility of motivating Vietnamese in the first place, the USIA began to absolve itself from PSYOP by the middle of 1966 in order to concentrate on what it had always considered its primary duty: securing good press. The agency had begun to feel squeezed by the increasing demands to produce in both areas throughout the year. The USIA contended it did not possess adequate resources to carry out both tasks.

In May 1966, Zorthian requested that the USIA purchase additional press to "maximize" the success of Chieu Hoi. The JUSPAO director called for a "considerably increased effort" from his agency, "especially in printing such materials as leaflets and posters." Zorthian wanted to double JUSPAO printing, but he found that the USIA was "reluctant" to approve of his proposal. It would necessitate "hiring additional Vietnamese and maybe bringing several more Americans into Viet-Nam." The USIA suggested that the JUSPAO's additional printing needs could be met by using the USIA plant at Manila. Further, the agency called on MACV to airlift the materials into Vietnam.[28] USIA director Leonard Marks presented the idea to Komer, who convinced MACV to handle transport duty.[29]

Later in 1966, the JUSPAO went to Ambassador Porter, who then went to Komer, to beseech the USIA for adequate funding for programming needs. The first time occurred in July, when Komer reported to Marks that Porter had "asked me to assist in obtaining a small but important increase in the JUSPAO budget." The amount of money he requested, "only $20,000," would enable the expansion of culture-drama teams to expand from eleven to fifteen. Komer noted that Zorthian had already secured $115,000 of the total program cost and asked whether the "USIA can allocate this minor amount of funds" to make up the difference.[30] Once again in October, Komer went back to Marks on behalf of the JUSPAO, indicating concern over the USIA proposed cutbacks to Chieu Hoi. He pleaded with Marks, "Appreciate your over-all problems but sure you will agree expanded Chieu Hoi has top priority."[31] The USIA's unwillingness to budge left Chieu Hoi "impeded by a lack of funds" for the remainder of the year.[32]

Marks finally placed the writing of his plans for the USIA on the wall in December in a correspondence with Komer regarding the first step on the way to CORDS, the creation of the Office of Operations (OCO) under ambassador Porter that served to unify the civilian agencies pursuing pacification. He wrote to Komer that he did not intend "to allow JUSPAO to involve itself deeply and visibly in the operational aspects of psywarfare." He insisted these elements of operation are "the function of the military." Marks explained his position:

> Indeed, USIA has become more deeply involved operationally than is proper, basically because the military needed assistance. With the Porter reorganization, the time is right for USIA to begin to hand back to the military those psywar

> functions in the field which are properly DOD's [Department of Defense] primary responsibility.[33]

Along with other changes, Marks called for replacing all thirty-four of the civilian Field Representatives that the JUSPAO utilized with military officers at the end of their current tours of duty. He stressed, "I have told the President and Secretaries McNamara and Rusk on many occasions of the inordinate drain which our staffing of JUSPAO has placed on USIA's limited resources." The point had been reached, Marks claimed, where the USIA "simply cannot continue to provide so many USIA officers for non-USIA type functions."[34]

Komer could not have agreed more with Marks. A week later he remarked to Ambassador Porter in Saigon that JUSPAO field representatives are "frankly not being fully employed in some cases." Moreover, their continued presence in Vietnam caused "a quite serious distortion of USIA's worldwide personnel structure."[35] Marks continued to push Zorthian to streamline the JUSPAO operation. He cabled Zorthian to suggest that the JUSPAO curtail its printing costs. Marks wrote, "Although leaflet operations are important to Chieu Hoi/surrender campaigns," it "appears here that disproportionate share of Psychological warfare/operations assets being devoted this medium which may or may not detract from others." Marks called on the JUSPAO to substitute leaflets with greater GVN participation. Prior experience demonstrated, he added, "what Vietnamese can do when motivated by top leaders."[36]

The USIA plan did not sit well with Zorthian. By the end of 1966, in what he dubbed "a rather lengthy communication" with Marks, he attributed its origin to the agency's "essential problem" of "manpower and funds." The JUSPAO director asserted that he believed the problem was solvable within the existing framework, "one way or another." If the plan had roots in philosophy, specifically "the JUSPAO concept of a functional, integrated approach to the whole problem of communications," then Zorthian implored Marks to "reconsider any retreat on the part of the Agency."[37]

It is not clear to what extent Zorthian feared, as his critics allege, the USIA plan would diminish his so-called communications empire. It is clear, though, that he believed the USIA possessed a vested interest in continuing to support the existing JUSPAO arrangement. He contended that "the hurt to the Agency" had peaked. Each new USIA officer assigned to Vietnam would return possessing experience "that should prove highly useful elsewhere." Zorthian also warned Marks against succumbing to the current climate of opinion in Washington that encouraged change for the sake of change and "groping for instant solutions to what are clearly difficult and complicated problems." Above all else, he stressed his belief that the "integrated arrangement" of the JUSPAO constituted "the most effective" tool for carrying out

PSYOP. Tearing apart the JUSPAO, Zorthian believed, would create an imbalance favoring the military. He reminded Marks, "It should be said that the Vietnamese to whom we are addressing ourselves have a deep-rooted suspicion of all things military."[38]

What eluded Zorthian, however, was the fundamental dynamic that drove Marks to continue pushing the change until it fully took effect in February 1968. By that point, the CORDS alternative under Komer firmly had taken hold; the USIA had reduced its staff in Vietnam, practically dissolving the JUSPAO; and Zorthian controlled only the Mission Press Center. While Zorthian perceived integration between the military and civil side as the key to winning hearts and minds, officials in Washington had come to consider it a time consuming distraction.

Vietnamese loyalty would come most efficiently, officials believed, through a military directed and executed pacification program. In the meantime, the USIA could set about doing the job it had been created to carry out: facilitating the approval elsewhere for American policies. Agency leaders recognized that their decidedly unsubtle attempt to integrate psychological warfare and information had failed because it diverted too much from the agency's primary mission. According to Marks, the JUSPAO concept risked "jeopardizing the effectiveness of USIA programs in other parts of the world where Viet-Nam needs to be explained and where understanding of or support for American policies needs to be assured."[39]

In April 1967, the USIA issued a new set of guidelines regarding its responsibilities in instances it deemed as "limited war and insurgency situations." According to its new regulations, it assigned the Department of Defense the responsibility for PSYOP. "USIA should direct or carry out such operations only in an emergency when specifically instructed to do so by the President." Any help it provided to a foreign government targeted by insurgents, the directives instructed, would be confined to a "definite time-table for its termination" and assigned "only to central information services." The USIA believed it had learned the limits of its ability to help "the modernization process to maturity" in Vietnam.[40] It willingly relinquished its responsibility for PSYOP in Vietnam to the nation's military, which eagerly reabsorbed it to deploy as a tactical weapon in the familiar style within the American Way of War.

Consequently, the Department of the Army published Field Manual 33-1 (FM 33-1) in June 1968 to supersede the PSYOP doctrine it issued in May 1965. FM 33-1 asserts that "the U.S. Army" has "the primary U.S. military role in PSYOP." It identifies several objectives for PSYOP, which include: "to gain, preserve and strengthen civilian support for the host government and its stability operations objectives," and "to establish and maintain a suitable image for the U.S. elements supporting the host country." Its emphasis, however, principally remained on offensive tactical PSYOP aimed at creating "dissension, dissatisfaction and desertion among insurgent forces."[41]

In 1969, the JUSPAO reported that its Chieu Hoi campaign for Tet, which sought "to exploit seasonal nostalgia and encourage defections among enemy troops," had become "the largest single psychological operation in Vietnam." Preparations for the campaign consumed almost six months and resulted in fifty-seven items "ranging from leaflets and posters to cartoon books and magazines." The JUSPAO cited the "excellent cooperation" that it received from the military, "which provided $94,000 for the purchase of person-to-person PSYOP items bearing Tet messages." This contribution helped to make the 1969 campaign the "most thoroughly planned to date" and one that resulted in a "higher volume of PSYOP activity than in any previous year."

As noted by the JUSPAO at the time, this campaign owed its success, at least in part, to the great extent in which the style of PSYOP employed had managed to complement the military offensive that simultaneously took place. American and Vietnamese forces "were strongly on the offensive" during this period and created "conditions deleterious to enemy morale." Citing the strong PSYOP support offered by military elements, the JUSPAO pointed out that "nearly all" of its "development work for the military is in the area of defection-inducement propaganda." It is "reasonable to conclude," the JUSPAO offered, that "allied military conditions created the pressure wherein enemy soldiers were more vulnerable to the stepped-up PSYOP campaign."[42]

NOTES

1. Joint State-USIA cable, October 4, 1965, *The Pentagon Papers,* Gravel edition, II: 543.
2. Report, Honolulu Meeting: Record of Conclusions and Decisions for Further Action, February 23, 1966, Honolulu Conference February 6–8, 1966, volume 2, box 44, NSF—NSC History, LBJL.
3. Rowan, *Breaking Barriers*; Sorensen, *The Word War*.
4. Sorensen, *The Word War*, 274.
5. Memo, for the President (Johnson), Weekly Report, February 15, 1966, "USIA—1966," box 135, Confidential File—Agency Reports—USIA, LBJL.
6. Memo, USIA Weekly Report for the President, February 14, 1967.
7. USIA Weekly Report, February 14, 1967, "USIA Volume 7, 1 of 2," box 75, NSF—Agency File—USIA, LBJL.
8. USIA Weekly Report, November 8, 1966, "USIA—1966," box 135, Confidential File—Agency Reports—USIA.
9. "Of Junkets and the USIA," *Time* magazine, August 26, 1966, 8.

10. Letter, to JUSPAO Director (Zorthian) from USIA Director (Marks), August 23, 1966, "Material on JUSPAO," box 4A39, Papers of Barry Zorthian, LBJL.

11. Report, Intelligence Note—232, Impact of Hanoi's "Peace Offensive" on U.S. Relations with Canada and Western Europe, March 22, 1967, "Vietnam Memos Volume 68, March 20–31, 1967," box 42, NSF Country File—Vietnam, LBJL.

12. Memo, Memorandum for the President, October 5, 1966, "Memos to the President July–December 1966," box 1, 2, NSF—Files of Robert Komer (emphasis in original).

13. Memo, Memorandum for the President, October 5, 1966.

14. Sheehan, *A Bright Shining Lie*, 655.

15. Letter, To AMEMB SAIGON for Zorthian from Komer, June 1, 1966, "Psychological Warfare," box 21, NSF—Komer-Leonhart File, LBJL.

16. Memo, Memorandum for the President, April 27, 1967, "Vietnam Memos (A) volume 70, April 25–May 8, 1967," box 43, NSF—Country File—Vietnam, LBJL.

17. Report, Pacification/RD Briefing, undated, "Vietnam Memos, volume 68, March 20–31, 1967," box 42, LBJL.

18. See *The Pentagon Papers,* Gravel edition, II: 515–89.

19. Letter, JUSPAO director Zorthian to USIA director Marks, December 21, 1966, "Material on JUSPAO," box 4A39, Papers of Barry Zorthian, LBJL.

20. Letter, JUSPAO director Zorthian to Executive Secretary Ad Hoc Committee on Army Psychological Operations, November 12, 1966, "Material on JUSPAO."

21. Memo, Comments Upon Completion of Assignment with JUSPAO, June 17, 1966, "Material on JUSPAO" (emphasis in original).

22. Memo, General J. F. Freund's Completion of Tour Memorandum, July 24, 1966, "Material on JUSPAO."

23. Report, JUSPAO Support for Revolutionary Development, July 25, 1966, "NSF—Komer-Leonhart File," box 21, Psy-War, LBJL.

24. Memo, from William Leonhart to President (Johnson) Report on Visit to Vietnam, May 17–29, 1966, May 30, 1966, "NSF—Komer-Leonhart File."

25. Letter, USIA Director Marks to JUSPAO Director Zorthian, August 23, 1966, "Material on JUSPAO," box 4A39, Papers of Barry Zorthian, LBJL.

26. Letter, USIA Director Marks to JUSPAO Director Zorthian, November 2, 1966, "Material on JUSPAO."

27. Memo, Holbrooke to Komer and Ambassador Leonhart, After JUSPAO, How? April 5, 1967, "Vietnam Memos volume 69, April 1–24, 1967," box 42, NSF Country File—Vietnam (parentheses in original).

28. Cable, Komer to Porter, May 7, 1966, "Backchannel Cables, Porter—Komer, May–July 1966," boxes 3, 4, NSF—Files of Robert W. Komer.

29. Memo, Komer to Marks, May 11, 1966, "JUSPAO," Papers of Leonard Marks, LBJL.

30. Memo, Komer to Marks, Increase in Village Psychological Operations in Vietnam in FY 1967, July 28, 1966, "Psychological Warfare," box 21, NSF Komer—Leonhart File.

31. Telegram, for Marks from Komer, October 29, 1966, "Psywar," box 8, NSF—Files of Robert Komer.

32. Report, Some Lessons Learned About Chieu Hoi in 1966, February 7, 1967, "Chieu Hoi (1)," box 2, NSF—Komer—Leonhart File.

33. Memo, for the President's Special Assistant Komer from USIA director Marks, USIA Support to Saigon Mission's New "Office of Operations," December 7, 1966, "Psychological Warfare," box 21, NSF—Komer-Leonhart File.

34. Memo, for the President's Special Assistant Komer, December 7, 1966, "Psychological Warfare."

35. Letter, to Ambassador Porter from Robert Komer, December 14, 1966, Psywar, box 8, NSF—Files of Robert Komer.

36. Cable, to Saigon from Marks, February 3, 1967, NSF—Files of Robert Komer.

37. Letter, to USIA director Marks from JUSPAO director Zorthian, December 21, 1966.

38. Letter, to USIA director Marks from JUSPAO director Zorthian, December 21, 1966.

39. Memo, for United States Under Secretary of State Nicholas Katzenbach from USIA Director Leonard Marks, August 1, 1967, "Material on JUSPAO," box 4A39, Papers of Barry Zorthian, LBJL.

40. Report, USIA Responsibilities in Limited War and Insurgency Situations, April 26, 1967.

41. Psychological Operations—U.S. Army Doctrine, June 1968, folder 10, box 4, Douglas Pike Collection, Unit 11—Monographs, VA-TTU.

42. Letter, from JUSPAO Saigon to USIA Washington: PSYOP Support, folder 9, box 13, Douglas Pike Collection, Unit 02—Military Operations, VA-TTU.

Chapter Eleven

From the People, For the People

OVERVIEW

This chapter is divided into four parts. The first, "Trick Lights and Soft Focus," demonstrates that, as the USIA lessened its commitment to the JUSPAO, it remained dedicated to promoting American policies in Southeast Asia. Such activity aroused criticism that the executive branch misused the USIA prompting congressional hearings and reform. Still, despite its insistence otherwise, American PSYOP had blurred the line between information and persuasion for much of the time it operated in Vietnam. The second section, "Lost in Translation," examines the assessments made by the JUSPAO of its work in Vietnam. According to agency findings, the JUSPAO achieved little success in communicating persuasively with Vietnamese.

This chapter's third section, "A Lack of Intelligence," raises the question of whether Americans took seriously the intelligence function of PSYOP. It demonstrates that PSYOP possessed few intelligence resources of its own and that it had little access to other streams of American intelligence. In Vietnam, PSYOP intelligence generally resulted only from extraordinary efforts made by individuals working outside the system. The chapter's final section, "By the People," identifies lessons offered by the use of American PSYOP in Vietnam. It also questions whether they have been adequately learned.

TRICK LIGHTS AND SOFT FOCUS

The USIA quickly resumed its endeavor to garner favorable impressions around the world for American policies and to increase the national standing in relation to the communist world. Its efforts to do so, however, raise more questions than can be answered. Even as the USIA moved to reduce its commitment to the JUSPAO, it dedicated itself to promoting first the Johnson administration then, under former president of CBS television Frank Shakespeare, the administration of Richard Milhous Nixon. In doing so, the USIA further blurred the line between soliciting support for the United States and furthering the political fortunes of particular presidential administrations.

The perception of any president's foreign policy is inseparable from that individual's prospects of staying in office. There is a fine line between responding to public opinion and attempting to shape it. The USIA expressed a decided interest in helping Johnson do the latter to increase his own advantage at the polls. In 1968, USIA director Marks, an old friend of Johnson, reminded the president that, in the 1960 election, "one of the issues which attracted considerable public attention was 'declining prestige of the U.S. abroad'" under the Republican administration of Eisenhower. He indicated the "possibility that a similar issue will be raised this year" and stressed the opposition "will not be able to justify any contention of this nature by reference to USIA material."

Marks warned Johnson that the Republican Coordinating Committee had issued a position paper criticizing "The lack of emphasis put on this activity by your Administration." He assured him, "there are no scientific indicators of this nature."[1] Actually, Johnson's administration had already taken steps to insure that Vietnam would not be a detriment to his political fortune. The JUSPAO had initiated a continuing sampling of Vietnamese throughout the countryside in 1966 in an effort to gauge their views about the RVN, the war, and the United States. According to Harry Latimer, former JUSPAO Chief of Plans, Policy and Research, this effort ended by an edict issued by the White House, "reportedly because of possible domestic political repercussions if the results got out."[2]

The Oval Office raised no objections, however, about the USIA promotion of Johnson's presidency. In October 1966, the agency held a private viewing of "A President's Country," a twenty-eight minute documentary "describing the making of a President in the elemental language of earth, sky, water and rock." The USIA paid Guggenheim Productions of Saint Louis, Missouri, eighty thousand dollars to produce the movie, which was narrated by the acclaimed Hollywood actor Gregory Peck. The agency had the film translated into five languages and distributed two hundred and sixty-five prints to ninety-eight countries. The movie, a "brooding saga full of memories and symbols," pays tribute to Johnson's Texas origins. The camera finds Johnson on his ranch wearing a broad-brimmed hat against a corral fence. Peck's narration adds:

> It can take a man a long time to travel from Texas to Washington, D.C., but when he goes, he carries with him the heritage that will shape his spirit and govern his judgments until the end of his days.[3]

the USIA was prohibited by law to release its products within the United States. Still, it made the "A President's Country" available for reporters and advised Johnson of the favorable reviews that appeared in the domestic press.

In addition to the filmmaking it carried out on Johnson's behalf, the USIA rededicated the JUSPAO to getting favorable stories out of Vietnam. In particular, the JUSPAO embarked on an "increased effort to cover Vietnam for third country audiences." It worked to strengthen the "program of backgrounders, talking papers, monographs and source material" for distribution to American officials around the world. Its objective was to provide them with the means "to do a more effective job in dealing with foreign public opinion." The renewed effort by the JUSPAO included a number of other initiatives. For example, the agency engaged in the distribution of materials "directly to non-American foreign correspondents," an increased output of "tailored stories and still photos," and the use of "publications, pamphlets, films, and radio programs to tell our story in other countries."[4]

In the end, Johnson's use of the USIA to project a favorable impression of his policies made little difference regarding the public's perception. On March 31, 1968, a short time after receiving the note from Marks concerning the Republican Coordinating Committee, the president announced he would not seek another term in office. The massive uprising that occurred in early 1968 in southern Vietnam, the Tet Offensive, brought the gap between official rhetoric and the actual conditions in Vietnam into sharp focus for the American public. The consequent backlash against American officials regarding Vietnam made manifest the warning about the nation's PSYOP offered by the *Washington Post* years before, when it suggested that such activities proved useful "in small doses during an emergency, as a doctor might prescribe drugs, but, taken too often and in too large doses, they might create addicts."[5]

Johnson's successors, Richard Milhous Nixon and Gerald Rudolph Ford, both republicans, continued to utilize the USIA as a public relations tool for their policies in Southeast Asia. Such use produced great controversy when it became public that the USIA directed its overseas radio programming through the Voice of America (VOA) to provide no coverage of student demonstrations in Phnom Penh on March 19, 1975, which called for the removal of the American-backed Cambodian dictator Lon Nol. Then, as a result of the USIA decision for VOA to downplay America's evacuation of its embassy in Saigon the following month, the United States Congress opened hearings into USIA activities. Representative Bella Savitsky Abzug, a democrat from New York who presided over the investigation, acknowledged that, "Following Vietnam and Watergate, the credibility of this Government and its leaders depends on the forthrightness of not only our message, but our messengers."[6]

The public pressure for reform that these proceedings helped generate led to Congress granting VOA an independent charter in 1976. It also prompted the James Earl Carter presidential administration to reorganize and rename the USIA in 1978 as the International Communication Agency (ICA). This change proved temporary, however, as President Ronald Wilson Reagan reconstituted and reinvigorated the USIA in 1982 for stepping up America's effort in the Cold War against the Soviet Union.[7]

A decade prior to the efforts made by Congress to reform the USIA, the agency's executive assistant deputy director Howard L. Chernoff addressed a New York audience in 1966 about the role played by the JUSPAO in America's "Other War" in Vietnam. He suggested, "True national images—like personal ones—cannot be created by trick lights or soft focus."[8] Nevertheless, this strategy is the one American officials adopted in Vietnam. In fact, there exists little evidence they ever considered using any other means. By the time they reached Vietnam, they had become thoroughly boxed in by the ideas they held about Vietnamese, about communists, and about themselves. Officials never managed to escape, or even to recognize, their entrapment. Nothing demonstrates this situation more vividly than the way officials conceived of, and then used, their PSYOP program in Vietnam.

Considering the prospect of holding national elections in the RVN, for example, in 1966, JUSPAO director Barry Zorthian submitted his thoughts to Ambassador William J. Porter. He observed, the "results of the elections are of crucial importance to our entire posture in Vietnam." He warned, though, that "if we adopt a passive attitude towards the present weakness of the government apparatus," then "we will place our entire effort here in jeopardy." Zorthian contended that America's investment of "a quarter of a million men and many billions of dollars justifies our insistence on full partnership." He proposed "endorsing publicly the whole process of democratic elections and then seeking to use our influence actively but quietly to encourage those whose election will be in our interest." American intervention should be "carefully synchronized and skillfully executed," Zorthian argued. He made it clear:

> But we should not be fuzzy to ourselves about our objectives. We should and we must have an active role in the next few months on the Vietnamese political scene if U.S. interests are to be safeguarded.[9]

LOST IN TRANSLATION

The JUSPAO director worried about the fuzziness in American minds. Nevertheless, it is the Vietnamese who remained hazy. According to a 1967 Nationwide Hamlet Survey that the JUSPAO conducted thirteen years after

Lansdale and his team of supporters from the USIA arrived in Vietnam, few Vietnamese had come to perceive members of the NLF as outsiders. Only 23% of those who responded to the survey considered RVN opposition as orchestrated by Communist China and Russia. Just 18% considered them "enemies of the people, traitors to the nation," while 5% even saw them as liberators. When asked to "describe in a few words what a Viet Cong is," the survey reported 30% of the respondents "merely say the Viet Cong are Vietnamese and 23% express no opinion at all."[10]

The JUSPAO had also not succeeded in convincing Vietnamese of the altruism of America's motives; only 40% approved of the United States decision to intervene while the rest either disapproved (40%) or expressed uncertainty (20%) about the American presence in their country. Most respondents cited colonialism and economic expansionism, as opposed to the eradication of communism, as the reasons for the continued conflict. The majority (72%) had little idea about how to end the war, but according to the survey, "whatever the suggestions offered, about three times as many want the Government of South Vietnam to 'do it' as want the U.S. to 'do it.'"[11]

By this point in time, the JUSPAO had been vigorously working in Vietnam for over a decade to promote the idea that communist insurgents constituted nothing more than terrorists that had betrayed their nation. By its own assessment, the agency had proven itself unable to sell this notion widely, even when only considering the views of those Vietnamese that Americans could consider as sufficiently pacified. The JUSPAO survey, carried out from July–September 1967, canvassed 122 hamlets in 15 provinces located through all four Corps regions in the RVN to interview 1,313 respondents. As the reports make clear, though, "the survey had to be restricted to localities which are reasonably secure and accessible."[12]

Consequently, Vietnamese satisfaction with the level of security among those interviewed ranged from 70 to 90% satisfaction. Still, despite America's conviction that security represented the key issue for Vietnamese, who only wished to be left alone and in peace, relatively few of the people that the JUSPAO interviewed testified to experiencing any new sense of happiness as a result of their heightened sense of safety. Between only 13% and 33% of the survey's respondents believed their lives had improved over the past year while 33% to 49% considered them as "worse than last year."[13]

While Americans secured few hearts in minds residing within the RVN, they fared no better in relaying their message to the DRV. By 1970, American officials concluded "The North Vietnamese leaders do not offer a responsive target audience with respect of direct psyops programs." United States credibility, they offered, "is simply too low." While "well conceived programs directed against the various target audiences of the NVN" might "harass and cause some reactions from the enemy," American PSYOP could expect to accomplish little more from further efforts. "Causing a change of NVN policy," they contended, "will result only from what action the NVN leaders infer from what they see happening in SVN and the USA." American planners concluded it "reasonable to say" that successful GVN programs and successful GVN activities in the SVN will be "the medium by which the NVN leaders will be most influenced."[14]

A LACK OF INTELLIGENCE

Americans failed to communicate persuasively in Vietnam because they invested little in learning about Vietnamese in the first place. A study completed in 1967 carried out by the 15th PSYOP Detachment, 7th PSYOP Group, concluded that PSYOP intelligence had attained only a "low level of effectiveness in Vietnam." It did not attribute the situation to a lack of intelligence that amounted to "over 1,500 written pages being issued every day." Much of this material, the report concluded, "while a by product of the primary collection effort of military intelligence information," represented "much information, which is useful input to psychological operations intelligence analysis."[15]

The American Way of War's style of PSYOP did not prioritize intelligence concerned with Vietnamese hearts and minds. PSYOP possessed minimal collection activities of its own which made it dependent on the "fallout" generated from other efforts. This situation proved problematic, as intelligence collectors possessed a limited capacity to respond to PSYOP needs. The Combined Military Interrogation Center (CMIC) in Tan Son Nhut, operated jointly by U.S. and RVN military personnel, and the National Interrogation Center in Saigon (NIC), run by the RVN, interrogated prisoners of war and some Chieu Hoi. Both operations offered great potential for PSYOP that went largely unfulfilled.

At the CMIC, interrogators "had no formal schooling in psychological operations." Personnel received their training at the United States Army Intelligence School in Fort Holabird, Maryland, which used *The Interrogator's Guide* as its basic text. Of the twelve military activities it lists on page one for the interrogator to coordinate, the guide identifies as first Order of Battle Personnel and last as PSYWAR personnel. Interrogators in training received ideas for PSYOP questions from the book, but they had to search for these examples, which are scattered about throughout the text. The suggested questions concern vague areas such as morale but neglect to address specific techniques such as leaflet and loudspeaker campaigns.[16]

This situation improved somewhat by the efforts made by the 4th PSYOP Group (formerly 6th PSYOP BN), Saigon, which had "no official access to military information facilities" due to MACV J-2 (Assistant Chief of Staff for Intelligence) procedures that discouraged "personal contact between psychological operations intelligence analysts and military intelligence collection agencies," preferring that all interaction occur by "papers sent through official channels." Utilizing "personal friendships and informal relationships," PSYOP interrogators gained access to the CMIC and the NIC, but the arrangement proved "a difficult way in which to operate."[17]

The dissemination of intelligence to PSYOP analysts also proved problematic in Vietnam. For example, Military Intelligence neglected to include both the 7th PSYOP Group and the 4th PSYOP Group on its direct distribution list and, despite the efforts made to correct this oversight, neither unit had made it onto the list by 1967. To compound its problem of receiving timely intelligence, the Psychological Operations Directorate (MACPD) insisted that all intelligence information directed to the 4th PSYOP Group route through it first before further distribution.

Obstacles other than red tape that impeded the circulation of intelligence materials included a lack of funding. PSYOP analysts managed to make only limited use of documents churned out of the Combined Document Exploitation Center (CDEC) located in Tan Son Nhut. The CDEC received copies of documents from all over Vietnam, which it translated, producing up to 120,000 pages per day. This immense CDEC output went into its Document Retrieval System that stored the material on film, indexed in detail according to the Intelligence Subject Code. The freely available system came to include all Intelligence Reports pertaining to Vietnam through the CONFIDENTIAL classification. It remained beyond the reach of PSYOP analysts, who, with the exception of those at the JUSPAO who benefited from a J-2 Liaison Officer that made daily visits to CDEC, lacked the 3M Reader Printers necessary to read the film.

This situation in field offices not only prevented the circulation of material, it rendered an unwieldy storage and retrieval system that made officials dependent on file cabinets. Even when they did secure access to CDEC documents, usually by visiting the center in person, its usefulness remained limited. Materials classified as SECRET or higher remained off limits to PSYOP intelligence except when personnel could arrange to view them through the "personal effort of a friend." In this case, though, the final product received a CLASSIFIED stamp and achieved only limited circulation.[18]

No factor hindered PSYOP intelligence more than personnel issues. Researchers and analysts numbered too few, and those who did the work possessed few qualifications. While the study conducted by the 7th PSYOP

Group in 1967 recommended that the PSYOP effort in Vietnam required almost four hundred individuals engaged in intelligence at all levels, it actually employed twenty-five. At the time of the report, six worked in the JUSPAO Planning Office, six sat the Psychological Desk in the Combined Intelligence Center, Vietnam, two operated as Psychological Officers for MACV J-2, two more resided in MACPD, and the 4th PSYOP Group (6th PSYOP Battalion) utilized six in its Research and Analysis Team. Because of this condition, the report's authors concluded that "those who are involved must exist from crisis to crisis, like firemen putting out fires." Moreover, they observed that these analysts lacked "adequate time" that would allow them to "do the more thorough and detailed studies that are needed." The evidence suggests that, in the case of the JUSPAO, personnel lacked the human resources to scrutinize up to ninety-five percent of the intelligence that arrived at its doors daily.[19]

PSYOP intelligence in Vietnam not only suffered from understaffing, it generated a product skewered toward the national level. All of the analysts it possessed worked in Saigon. None operated at the Corps, Division, or Sector level. This situation resulted in generalized products of marginal utility for Sector Advisors, Province Representatives, and PSYOP teams providing tactical support. The JUSPAO pre-tested and evaluated products using Vietnamese, for example, but its interviewing and surveying remained confined to Saigon and the safe areas that surrounded it. The results it obtained from questioning Hoi Chanh at Chieu Hoi centers, while often useful locally for "confirming or identifying trends," often did not prove timely enough.

Instead, PSYOP officials in the field often found that "locally produced leaflets were superior to those produced at the Saigon level because they are based on first hand information and tactical intelligence, and are designed with specific targets in mind." For example, one Provincial PSYOP Officer reported that a "nationally generated TET campaign did not really have much pertinence in Darlac [Dac Lac] Province." During the campaign, over four million leaflets rained from the sky over Darlac Province. The official pointed out, though, "TET is a celebration of the ethnic Vietnamese who are concentrated in the lowland coastal areas of Vietnam, and only the towns of the highland area." The leaflets, he continued, proved "valueless" when used against the non-ethnic Vietnamese tribes of Darlac (E De), "they do not celebrate TET."[20] As late as 1969, the JUSPAO conceded, "There is a need for established procedures for testing PSYOP material in Viet Nam."[21]

PSYOP intelligence in Vietnam resulted from efforts made by individuals who possessed few qualifications to perform their assigned jobs. "In many cases," according to the study conducted by the 7th PSYOP Group, individuals held Intelligence Research and Analysis positions "who have had no experience in psychological operations, intelligence, or related fields in the social sciences." Moreover, none had received "Southeast Asia studies training." Without the proper background, PSYOP intelligence officials experienced difficulty relating to the larger intelligence community. They simply did not "fully understand" its "ability to serve them." Consequently, a cycle of frustration operated, where "products given to PSYOPS" did not fulfill its needs and PSYOP staffs did not "go to the intelligence community for specific help to meet specific requirements as often as they might."[22]

They also experienced difficulties in connecting with Vietnamese. Like most of their colleagues engaged in PSYOP work, they possessed little specialized training relevant to their jobs. The JUSPAO employed nobody who possessed an advanced degree in psychology and had no individuals working on field teams or PSYOP battalions who held an advanced degree in any related social science. One study found that, over a three year period, no PSYOP officer in the military's field forces and no civilian CORDS Regional Directors or their subordinate assistant province advisors for PSYOP, had any graduate training in psychology.

The same held true, according to the report, for the 4th PSYOP Group Headquarters. Based on its analysis of "everyday operations of psyop units plus the results of a survey of the qualifications, background, training and experience of personnel involved in psyop at all levels" in Vietnam, from "individual action" to "the highest level of policy formulation," the study asserted "that there is little or no input of competent professional knowledge of dynamic applied psychology to most aspects of PSYOPS." It concluded that labeling the American effort as psychological operations proved "misleading even if interpreted very loosely."[23]

FOR THE PEOPLE

The JUSPAO operation's rhetorical promise offered the potential for enabling a genuine dialogue to develop between Americans and others regarding Vietnam. Instead, it came to represent a colossal missed opportunity on the part of the United States to utilize psychological operations in an imaginary and far-sighted way that might have injected its overall policy with both the sensitivity and the flexibility it otherwise lacked. American officials had convinced themselves prior to arriving in Vietnam that their objectivity would render their nation's noble intentions apparent to all. When the approval they anticipated did not arrive, they set about to acquire it by using undemocratic means that they failed to detect would stand in stark contrast to the ends they purportedly sought. As the criticism mounted, they identified scapegoats lurking everywhere except for in the very places where they sat to draw up policy.

Even as they professed the need to win hearts and minds, officials made decisions that sacrificed gaining heartfelt indigenous support for American policies. Their insistence on pacification programs that emphasized security first failed to appreciate how these measures would resonate among a people who had been historically oppressed by outsiders. Officials confused the appearance of stability with progress. They did so, in part, because of the political pressures they felt from critics both at home and abroad. They did so also, though, because they never wavered from the idea that communism was something to be eradicated in a place like Vietnam. They refused to place their faith in the Vietnamese to grasp for themselves the ideals of freedom and democracy that Americans espoused.

Eager for results, American officials re-conceptualized PSYOP in Vietnam to consist of "all means of communication." In doing so, they combined two missions of PSYOP that Americans historically had managed to keep distinct from one another, information and persuasion, under the singular roof that the JUSPAO now provided. The new arrangement did little to reconcile the increasingly different views that military and civilian officials held from one another regarding the use of PSYOP. Military commanders clung to the notion that PSYOP, if useful at all, worked most effectively as a force multiplier by increasing the number of enemy surrenders and defections. Meanwhile, political leaders sought to employ PSYOP as a public relations tool to promote policies. Few individuals in either camp considered utilizing PSYOP to inform Americans about the Vietnamese hearts and minds that they coveted.

BY THE PEOPLE

In 1968, United States Army Colonel William F. Johnston labeled American PSYOP a "neglected deterrent." Johnston offered that, in the future, "insurgency threatened governments can be expected to heed U.S. advice if it is in the form of lessons learned at great cost to us in Vietnam." One "very important lesson" that Johnston identified called for the "establishment of a two-way communications system" that makes certain that "selected grievances and good ideas get to the psyops planners and decision makers at the top level." Only after "the true needs" of the people are "ascertained" and "transmitted to the top echelons" can there be "hope and opportunity" for the creation of appropriate policy. "Top level interest and support" must translate into "meaningful terms for people at the grass roots level."

According to Johnston, "the most important lesson" for PSYOP constituted the importance of "face-to-face communication." Officials "must recognize that a face-to-face psyops capability must be developed from the people, by the people, and for the people of a particular language group, clan, tribe, or area of the country." Johnston contended that "this is not something that can be 'Made in the U.S.A.' and exported." Instead, it is a process that requires "fully qualified and professional psyops experts experienced in the language and in the thinking of foreign people" who are able to cooperate fully with local populations. Such a PSYOP program, Johnston argued, would represent "one of the cheapest and best security investments" that America could make.[24]

While one would be hard pressed to disagree with Johnston's insights, it is difficult to discern whether American officials have shared in them. Amidst much controversy, in 2003, the Pentagon shut down its Office of Strategic Influence due to press reports that it was "mulling a scheme to plant false news items with foreign journalists to sway opinion overseas."[25] Moreover, continuing concerns about American military commanders blurring the line between the current information doctrine of Public Affairs (PA) and the persuasive doctrine of Information Operations (IO) led former chairman of the United States Chiefs of Staff General Richard B. Myers, to "issue a letter directing the military's top brass to keep PA distinct from deception and influence functions."[26]

Other contemporary efforts to win hearts and minds have also aroused critics. Al-Hurra (The Free One), a twenty-four hour Arab-language news station, was initiated by the George Walker Bush administration in 2004. Officials hoped the station, run by the Broadcasting Board of Governors, which also directs the Voice of America, would "help stem anti-Americanism fueled by the war on terrorism, the occupation of Iraq, and U.S. support for Israel." The Muslim Brotherhood quickly condemned the initiative, however, as "part of the American media and cultural invasion of our region." Rami G. Khouri, executive editor of Lebanon's *Daily Star*, predicted Al-Hurra would "exacerbate the gap between Americans and Arabs, rather than close it." Arab defenders of the station, such as Samiha Dahroug, head of Egypt's Nile News Channel, suggested that, in the end, the ability of the United States to win hearts and minds is dependent on its policies. "America is judged," she observed, "by how it conducts itself in the world."[27]

NOTES

1. Report, Biweekly Report, for the President (Johnson) from USIA Director (Marks), February 28, 1968, "USIA—1968," box 135, Confidential File—Agency Reports—USIA, LBJL.
2. Latimer, *US Psychological Operartions,* 79.

3. Memo, for the President, November 7, 1966, "USIA 1966," box 135, Confidential File—Agency Reports—USIA, LBJL.

4. Report, to Members of the Mission Council, August 7, 1967, "Vietnam 7E (1) a Public Relations Activities, September–October 1967," box 99, NSF—Country File—Vietnam, LBJL.

5. Quoted in Barrett, "PSYOP, What is it? What Should We Do About It?" *The PSYOP Digest*, 25.

6. House Subcommittee of the Committee on Government Operations, 94th Congress, 1st Session, May 20, 1975, Washington, D. C.: U.S. Government Printing Office, 1.

7. President William Jefferson Clinton dissolved the USIA in September 1999.

8. Article, "'The Other War in Viet-Nam,'" *USIA Correspondent,* March 1966, "USIA Correspondent 1965–1967," box 14, Papers of Leonard Marks, LBJL.

9. Memo, Zorthian to Porter, May 6, 1966, "Material on JUSPAO," box 4A39, Papers of Barry Zorthian, LBJL.

10. Report, Nationwide Hamlet Survey (Fourth Interim Summary Report—IV Corps), December 13, 1967, NSF—Country File—Vietnam, box 59, "Revolutionary Development Program 2, March–December 1967," LBJL.

11. Report, Nationwide Hamlet Survey, December 13, 1967.

12. Report, Nationwide Hamlet Survey (Second Interim Summary Report—II Corps), November 1, 1967, NSF—Country File—Vietnam, box 59, "Revolutionary Development Program 2, March–December 1967," LBJL.

13. Report, Nationwide Hamlet Survey (Fourth Interim Summary Report—IV Corps), December 13, 1967.

14. The North Vietnamese Leaders Do Not Offer a Responsive Target Audience with Respect to Direct Psyops Programs, June 1970, Folder 01, Box 17, Douglas Pike Collection: Unit 01—Assessment and Strategy, VA-TTU.

15. Report on Psychological Operations Intelligence in Vietnam, November 17, 1967, Folder 01, Box 14, Douglas Pike Collection: Unit 03—Insurgency Warfare, VA-TTU.

16. Report on Psychological Operations Intelligence in Vietnam, November 17, 1967.

17. Report on Psychological Operations Intelligence in Vietnam, November 17, 1967.

18. Report on Psychological Operations Intelligence in Vietnam, November 17, 1967.

19. Report on Psychological Operations Intelligence in Vietnam, November 17, 1967.

20. Report on Psychological Operations Intelligence in Vietnam, November 17, 1967.

21. JUSPAO PSYOP Circular Number 9: Testing of PSYOP Material in Viet-Nam, January 14, 1969, Folder 01, Box 17, Douglas Pike Collection: Unit 01—Assessment and Strategy, VA-TTU.

22. Report on Psychological Operations Intelligence in Vietnam, November 17, 1967.

23. Volume 1 Final Technical Report Psychological Operations Studies—Vietnam, May 25, 1971, Folder 01, Box 15, Douglas Pike Collection: Unit 03—Insurgency Warfare, VA-TTU.

24. Neglected Deterrent: Psychological Operations in "Liberation Wars," 1968, folder 12, box 15, Douglas Pike Collection: Unit 03—Insurgency Warfare, VA-TTU.

25. Article, "On the P.R. Battlefield," *Time* magazine, June 13, 2005, 13.

26. Scanlon, "In Defense of Military Public Affairs," 95.

27. Article, "Aimed at Arabs, U.S.-backed TV station draws Arab scorn," *The Philadelphia Inquirer*, February 14, 2004, 2.

Bibliography

ARCHIVE AND MANUSCRIPT COLLECTIONS

Administration History, (USIA), Lyndon Baines Johnson Library, Austin, Texas.
Barry Zorthian Papers, Lyndon Baines Johnson Library, Austin, Texas.
BDM Corporation Study: A Study of Strategic Lessons, Lyndon Baines Johnson Library, Austin, Texas.
Carl Nelson Collection, The Vietnam Archive, Texas Tech University, Lubbock, Texas.
Confidential File, (USIA Reports), Lyndon Baines Johnson Library, Austin, Texas.
Douglas Pike Collection, The Vietnam Archive, Texas Tech University, Lubbock, Texas.
Fred Panzer Office File, Lyndon Baines Johnson Library, Austin, Texas.
Leonard H. Marks Papers, Lyndon Baines Johnson Library, Austin, Texas.
Meeting Notes File, Lyndon Baines Johnson Library, Austin, Texas.
National Security File, (Agency File, USIA), Lyndon Baines Johnson Library, Austin, Texas.
National Security File, (Country File, Vietnam), Lyndon Baines Johnson Library, Austin, Texas.
National Security File, (Komer File), Lyndon Baines Johnson Library, Austin, Texas.
National Security File, (Komer-Leonhart File), Lyndon Baines Johnson Library, Austin, Texas.
National Security File, (National Security Action Memorandum), Lyndon Baines Johnson Library, Austin, Texas.
Oral History Collection, (Transcripts), Lyndon Baines Johnson Library, Austin, Texas.
Records of the Executive Office of the President and Independent Agencies, USIA Wireless File, Lyndon Baines Johnson Library, Austin, Texas.
Richard M. Lee Papers, (Counterinsurgency File), United States Army Military History Institute, Carlisle, Pennsylvania.
Robert W. Komer File, Lyndon Baines Johnson Library, Austin, Texas.

Senior Officer Oral History Program, (Transcripts), United States Army Military History Institute, Carlisle, Pennsylvania.
USIA Correspondent, 1965–1967, Lyndon Baines Johnson Library, Austin, Texas.
William J. Donovan Papers, United States Army Military History Institute, Carlisle, Pennsylvania.

PUBLISHED DOCUMENTS

Declassified Documents Reference System. Woodbridge, Connecticut: Primary Source Media, 1993. Microfilm.
Joint Chiefs of Staff. *Records of the Joint Chiefs of Staff, Part 1: 1942–1945, Strategic Issues*. Frederick, Maryland: University Publications of America, Inc., 1983. Microfilm.
———. *Records of the Joint Chiefs of Staff, Part 1: 1942–1945, The Pacific Theater*. Frederick, Maryland: University Publications of America, Inc., 1981. Microfilm.
The Fifteenth Report of the United States Advisory Commission on Information. Washington, D.C.: Government Printing Office, 1960.
The Pentagon Papers: The Defense Department History of United States Decisionmaking in Vietnam. Ed. Noam Chomsky and Howard Zinn. The Senator Gravel edition. 5 vols. Boston: Beacon Press, 1971–72.
The Thirteenth Report of the United States Advisory Commission on Information. Washington, D.C.: United States Government Printing Office, 1958.
United States Congress. House of Representatives. *United States of America Congressional Record: Proceedings and Debates of the 89th Congress, 2nd Session*. Washington, D.C.: Government Printing Office, 1966.
———. *National War Agencies Appropriations Bill, Hearings before a Subcommittee of the House Committee on Appropriations*. 78th Congress, 1st session. Washington, D.C.: Government Printing Office, 1944.
———. *Public Diplomacy and the Future, Hearings before the Subcommittee on International Operations of the Committee on International Relations*. 95th Congress, 2nd session. Washington, D.C.: Government Printing Office, 1977.
United States Department of State. *Foreign Relations of the United States: 1952–1954, National Security Affairs*. Washington, D.C.: Government Printing Office, 1984.
———. *Foreign Relations of the United States: 1961–1963, Vietnam*. 4 vols. Washington, D.C.: Government Printing Office, 1988–1991.
———. *Foreign Relations of the United States: 1964–1968, Vietnam*. 7 vols. Washington, D.C.: Government Printing Office, 1992–2003.

NEWSPAPERS AND PERIODICALS

"Aimed at Arabs, U.S.-backed TV station draws scorn." *The Philadelphia Inquirer*. February 14, 2004, 2.

Apple, R. W. "Saigon U.S.I.A. Hides Critical Books." *New York Times*. November 13, 1966, 2.
"Donovan is Named Information Head." *New York Times*. July 12, 1941, 1.
"Facts of Life in Vietnam." *Time* magazine. May 7, 1965, 13.
Fishel, Wesley R. "We Just Naturally Upstage Vietnamese." *The Washington Post*. December 27, 1964, E4.
Grose, Peter. "Score Vietnam Inertia." *New York Times*. April 20, 1964, 1.
"Need Nielsen to Count Vietcong Eavesdroppers on Ky's ETV Weapon." *Variety*. December 14, 1966, 1.
"Of Junkets and the USIA." *Time* magazine. August 26, 1966, 8.
"On the P.R. Battlefield." *Time* magazine. June 13, 2005, 13.
"OSS Plan to Fox Japs with Ghost Foxes Recalled." *Sunday Tribune*. March 21, 1948, 1.
"Popularity v. Power." *Time* magazine. November 7, 1960, 12.
"Psywar." *Time* magazine. October 29, 1965, 36.
Rhem, Kathleen T. "Burning Afghan Bodies Resulted from Poor Judgement, Not War Crimes." *American Forces Information Service*. November 29, 2005.
"Rumsfeld: U.S. Failing in Battle of Ideas." *The Philadelphia Inquirer*. March 24, 2996, 1.
"Ships Must Get Goods to Britain to Assure our Aid, says Donovan." *New York Times*. March 27, 1941, 1.
"Still Short in Iraq." *Time* magazine. March 14, 2005, 13.
"Truth or Propaganda." *New York Times*. April 23, 1965, 34.
"Wild Bill's Brain Bureau Set Up." *New York Times*. September 21, 1941, 1.

BOOKS

Ahern, Thomas L. *CIA and Rural Pacification in South Vietnam*. Center for the Study of Intelligence, undated.
Alexandre, Laurien. *The Voice of America: From Détente to the Reagan Doctrine*. Norwood, New Jersey: Ablex Publishing Corporation, 1988.
Ambrose, Stephen E., and Richard H. Immerman. *Ike's Spies: Eisenhower and the Espionage Establishment*. Jackson: University Press of Mississippi, 1981.
Anderson, Charles R. *Vietnam: The Other War*. Novato, California: Presidio Press, 1982.
Asprey, Robert B. *War in the Shadows: The Guerrilla in History*. Garden City, New York: Doubleday and Company, Inc., 1975.
Baritz, Loren. *Backfire: A History of How American Culture Led Us into Vietnam and Made Us Fight the Way We Did*. New York: Ballantine Books, 1985.
Barrett, Edward W. *Truth Is Our Weapon*. New York: Funk and Wagnalls, 1953.
Bergerud, Eric. *Red Thunder, Tropic Lightning: The World of a Combat Division in Vietnam*. Boulder, Colorado: Westview Press, 1993.
——. *The Dynamics of Defeat: The Vietnam War in Hau Nghia Province*. Boulder, Colorado: Westview Press, 1991.

Bidwell, Bruce. *History of the Military Intelligence Division of the War Department General Staff: 1775–1941*. Frederick, Maryland: University Publications of America, 1986.

Blaufarb, Douglas. *The Counterinsurgency Era*. New York: Free Press, 1977.

Bogart, Leo. *Premises for Propaganda: The United States Information Agency's Operating Assumptions in the Cold War*. New York: The Free Press, 1976.

Bohannan, Charles, and Napoleon Valeriano. *Counterguerilla Operations: The Philippine Experience*. New York: Frederick A. Praeger, 1962.

Bowie, Robert R., and Richard H. Immerman. *Waging Peace: How Eisenhower Shaped an Enduring Cold War Strategy*. Oxford: Oxford University Press, 1998.

Bowman, John S., ed. *The Vietnam War: An Almanac*. New York: World Almanac Publications, 1985.

Boyer, Paul. *By the Bomb's Early Light: American Thought and Culture at the Dawn of the Atomic Age*. New York: Pantheon, 1995.

Bradley, Mark Philip. *Imagining Vietnam and America: The Making of Postcolonial Vietnam, 1919–1950*. Chapel Hill and London: The University of North Carolina Press, 2000.

Brimmel, J. H. *Communism in Southeast Asia*. London: Royal Institute of International Affairs, 1958.

Brown, Anthony Cave. *The Last Hero: Wild Bill Donovan*. New York: Times Books, 1982.

Buntz, George G. *Allied Propaganda and the Collapse of the German Empire in 1918*. Stanford: Stanford University Press, 1938.

Burlingame, Roger. *Don't Let Them Scare You: The Life and Times of Elmer Davis*. Philadelphia: J. B. Lippincott Co., 1961.

Cable, Larry E. *Conflict of Myths: The Development of American Counterinsurgency Doctrine and the Vietnam War*. New York: New York University Press, 1986.

Carroll, Wallace. *Persuade or Perish*. Boston: Houghton Mifflin, 1948.

Chambers, John Whiteday, II; and G. Kurt Piehler, eds. *Major Problems in American Military History*. Boston and New York: Houghton Mifflin Company, 1999.

Chandler, Robert W. *War of Ideas: The U.S. Propaganda Campaign in Vietnam*. Boulder, Colorado: Westview Press, 1981.

Cohen Warren I., and Nancy Bernkopf Tucker, eds. *Lyndon Johnson Confronts the World: American Foreign Policy, 1963–1968*. Cambridge: Cambridge University Press, 1994.

Cooper, Chester. *The Lost Crusade: America in Vietnam*. New York: Dodd, Mead, 1970.

Costigliola, Frank. *Awkward Dominion: American Political, Economic, and Cultural Relations with Europe, 1919–1933*. New York: Hill and Wang, 1982.

Creel, George. *How We Advertised America*. New York: Harper & Brothers Publishers, 1920.

Currey, Cincinnatus. *Self-Destruction: The Disintegration and Decay of the United States Army During the Vietnam Era*. New York: W.W. Norton, 1981.

Daugherty, William E., ed. *A Psychological Warfare Casebook*. Baltimore, Maryland: Johns Hopkins Press, 1958.

Davidson, Phillip B. *Vietnam At War: The History: 1946–1975*. New York and Oxford: Oxford University Press, 1988.

Dean, Vera Micheles, and Harry D. Harootunian, eds. *West and Non-West: New Perspectives*. New York: Holt, Rinehart and Winston, Inc., 1963.

Dower, John W. *War Without Mercy: Race and Power in the Pacific War*. New York: Pantheon Books, 1986.

Duncanson, Dennis J., Richard A. Yudkin, and Barry Zorthian. *Lessons of Vietnam: Three Interpretive Essays*. New York: American-Asian Educational Exchange, 1971.

Dunlop, Richard. *Donovan: America's Master Spy*. New York: Rand McNally, 1982.

Eisenhower, Milton S. *The President Is Calling*. Garden City, New Jersey: Doubleday & Company, Inc., 1974.

Elder, Robert. *The Information Machine: The United States Information Agency and American Foreign Policy*. New York: Syracuse University Press, 1968.

Fabian, Johannes. *Time and the Other: How Anthropology Makes Its Object*. New York: Columbia University Press, 1983.

Fall, Bernard. *The Two Vietnams: A Political and Military Analysis*. New York: Praeger, 1964.

———. *The Viet-Minh Regime: Government and Administration in the Democratic Republic of Vietnam*. New York: Secretariat of the Institute of Pacific Relations, 1956.

Fineman, Daniel. *A Special Relationship: The United States and Military Government in Thailand, 1947–1958*. Honolulu: University of Hawaii Press, 1997.

Fitzgerald, Frances. *Fire in the Lake: The Vietnamese and the Americans in Vietnam*. Boston: Little, Brown & Co., 1972.

Ford, Corey. *Donovan of OSS*. Boston: Little, Brown and Company, 1970.

Fyne, Robert. *The Hollywood Propaganda of World War II*. Metuchen, New Jersey: The Scarecrow Press, Inc., 1994.

Gardner, Lloyd C., and Marilyn B. Young, eds. *Iraq and the Lessons of Vietnam: Or, How Not To Learn From the Past*. New York and London: The New Press, 2007.

Gilmore, Allison B. *You Can't Fight Tanks with Bayonets: Psychological Warfare against the Japanese Army in the Southwest Pacific*. Lincoln and London: University of Nebraska Press, 1998.

Green, Fitzhugh. *American Propaganda Abroad*. New York: Hippocrene Books, 1988.

Hallin, Daniel. *The "Uncensored War": The Media and Vietnam*. New York: Oxford University Press, 1986.

Hammer, Ellen. *The Struggle for Indochina, 1940–1955*. Stanford: Stanford University Press, 1955.

Hammond, William M. *Public Affairs: The Military and the Media, 1962–1968. United States Army in Vietnam*, edited by David F. Trask. Washington, D.C.: Center of Military History, United States Army, 1988.

Henderson, John W. *The United States Information Agency*. New York: Frederick A. Praeger, Publishers, 1969.

Herring, George C. *LBJ and Vietnam: A Different Kind of War*. Austin: University of Texas Press, 1994.

Hersh, Burton. *The American Elite and the Origins of the CIA*. New York: Charles Scribner's Sons, 1992.

Hixson, Walter L. *Parting the Curtain: Propaganda, Culture, and the Cold War, 1945–1961*. New York: St. Martin's Griffin, 1997, 1998.

Hogan, Michael J., and Thomas G. Patterson, eds. *Explaining the History of American Foreign Relations*. Cambridge: Cambridge University Press, 1991.

Hogan, Michael J., ed. *America in the World: The Historiography of American Foreign Relations Since 1941*. Cambridge: Cambridge University Press, 1995.

Hunt, Michael. *Ideology and U.S. Foreign Policy*. New Haven: Yale University Press, 1987.

Hunt, Richard A. *Pacification: The American Struggle for Vietnam's Hearts and Minds*. Boulder, San Francisco, and Oxford: Westview Press, 1995.

Jacobs, Seth. *America's Miracle Man in Vietnam: Ngo Dinh Diem, Religion, Race and U.S. Intervention in Southeast Asia, 1950–1957*. Durham, North Carolina: Duke University Press, 2004.

Jacobson, Matthew Frye. *Barbarian Virtues: The United States Encounters Foreign Peoples at Home and Abroad, 1876–1917*. New York: Hill and Wang, 2000.

Jamieson, Neil L. *Understanding Vietnam*. Berkeley and Los Angeles: University of California Press, 1993.

Janowitz, Morris. *The Professional Soldier: A Social and Political Portrait*. New York: Free Press, 1960.

Jefferys-Jones, Rhodri. *The CIA and American Democracy*. New Haven: Yale University Press, 1989.

Jongsuk Chay, ed., *Culture and International Relations*. New York, Westport, Connecticut, and London: Praeger Publishers, 1990.

Kahin, George McT. *Intervention: How America Became Involved in Vietnam*. New York: Alfred A. Knopf, 1986.

Kahn, E. J., Jr. *The China Hands: America's Foreign Service Officers and What Befell Them*. New York: Penguin Books, 1976.

Kaplan, Amy, and Donald Pease, eds. *Cultures of United States Imperialism*. Durham, North Carolina: Duke University Press, 1993.

Karnow, Stanley. *Vietnam: A History*. New York: The Viking Press, 1983.

Karsten, Peter. *The Naval Aristocracy: The Golden Age of Annapolis and the Emergence of Modern American Navalism*. New York: Free Press, 1972.

Kearns, Doris. *Lyndon Johnson and the American Dream*. New York: Harper & Row, 1977.

Kerkvliet, Benedict J. *The Huk Rebellion: A Study of Peasant Revolt in the Philippines*. Quezon City, Philippines: New Day Publishers, 1979.

Knightley, Phillip. *The First Casualty: From the Crimea to Vietnam: The War Correspondent as Hero, Propagandist, and Myth Maker*. New York and London: Harcourt Brace Jovanovich, 1975.

Kolko, Gabriel. *Anatomy of a War, Vietnam, the United States and the Modern Historical Experience*. New York: Pantheon Books, 1986.

Komer, Robert W. *Bureaucracy Does Its Thing: Institutional Constraints on U.S.-G.V.N. Performance in Vietnam*. Santa Monica, California: Rand Corporation, 1972.

Krepinevich, Andrew F., Jr. *The Army and Vietnam*. Baltimore, Maryland: The Johns Hopkins University Press, 1986.

Kuisel, Richard. *Seducing the French: The Dilemma of Americanization*. Berkeley, Los Angeles, and London: University of California Press, 1993.

Kutler, Stanley, ed. *Encyclopedia of the Vietnam War*. New York: Scribner's, 1996.

Latham, Michael E. *Modernization as Ideology: American Social Science and "Nation Building" in the Kennedy Era*. Chapel Hill: University of North Carolina Press, 2000.

Latimer, Harry D. *US Psychological Operations in Vietnam*. Providence, Rhode Island: Brown University, 1973.

Lerner, Daniel. *Psychological Warfare Against Nazi Germany: The Sykewar Campaign, D-Day to VE-Day*. Cambridge: The M.I.T. Press, 1971.

May, Elaine Tyler. *Homeward Bound: American Families in the Cold War Era*. New York: BasicBooks, 1988.

May, Lary, ed. *Recasting America: Culture and Politics in the Age of Cold War*. Chicago and London: The University of Chicago Press, 1989.

McAlister, John T., Jr., and Paul Mas. *The Vietnamese and Their Revolution*. New York, Evanston, and London: Harper & Ron Publishers, 1970.

McClintock, Anne. *Imperial Leather: Race, Gender, and Sexuality in the Colonial Context*. New York: Routledge, 1995.

McClintock, Michael. *Instrument of Statecraft: U.S. Guerilla Warfare, Counterinsurgency and Counterterrorism, 1940–1990*. New York: Pantheon Books, 1992.

McLeod, Mark W., and Nguyen Thi Dieu. *Culture and Customs of Vietnam*. Culture and Customs of Asia series, edited by Hanchao Lu. Westport, Connecticut, and London: Greenwood Press, 2001.

Mecklin, John. *Mission in Torment: An Intimate Account of the U.S. Role in Vietnam*. New York: Doubleday, 1965.

Melling, Phil, and Jon Roper, eds. *America, France and Vietnam: Cultural History and Ideas of Conflict*. Brookfield, Vermont: Gower Publishing Company, 1991.

Mickelson, Sig. *America's Other Voice: The Story of Radio Free Europe and Radio Liberty*. New York: Praeger, 1983.

Mock, James R., and Cedric Larson. *Words that Won the War: The Story of the Committee on Public Information, 1917–1919*. New York: Russell and Russell, 1939. Reprint, 1968.

Moskos, Charles C., Jr. *The American Enlisted Man*. New York: Russell Sage, 1970.

Nadel, Alan. *Containment Culture: American Narratives, Postmodernism, and the Atomic Age*. Durham, North Carolina: Duke University Press, 1995.

Nelson, Michael. *War of the Black Heavens: The Battles of Western Broadcasting in the Cold War*. Syracuse: Syracuse University Press, 1997.

Osgood, Kenneth. *Total Cold War: Eisenhower's Secret Propaganda Battle at Home and Abroad*. Lawrence, Kansas: University Press of Kansas, 2006.

Oshinsky, David M. *A Conspiracy So Immense: The World of Joe McCarthy*. New York: The Free Press, 1983.

Page, Caroline. *U.S. Official Propaganda During the Vietnam War, 1965–1973: The Limits of Persuasion*. London and New York: Leicester University Press, 1996.

Paterson, Thomas G., ed. *Kennedy's Quest for Victory: American Foreign Policy, 1961–1963*. Oxford: Oxford University Press, 1989.

Peterson, Michael E. *The Combined Action Platoons: The U.S. Marines' Other War in Vietnam*. Westport, Connecticut: Praeger, 1989.

Pirsein, Robert William. *The Voice of America: An History of the International Broadcasting Activities of the United States Government, 1940–1962*. New York: Arno Press, 1979.

Pollock, Daniel C., Project Director. Ronald De McLaurin, Carl F. Rosenthal, Sarah A. Skillings, and others, editors. *The Art and Science of Psychological Operations: Case Studies of Military Application*. 2 volumes. Washington, D.C.: American Institutes for Research, 1976.

Puddington, Arch. *Broadcasting Freedom: The Cold War Triumph of Radio Free Europe and Radio Liberty*. Lexington: The University Press of Kentucky, 2000.

Race, Jeffrey. *War Comes to Long An*. Berkeley and Los Angeles: University of California Press, 1973.

Roosevelt, Kermit. *War Report of the OSS*. 2 vols. New York: Walker and Company, 1976.

Rosenberg, Emily S. *Spreading the American Dream: American Economic and Cultural Expansion, 1890–1945*. New York: Hill and Wang, 1982.

Rostow, W. W. *The Stages of Economic Growth; A Non-Communist Manifesto*. Cambridge: Cambridge University Press, 1960.

Rowan, Carl T. *Breaking Barriers: A Memoir*. Boston, Toronto, London: Little, Brown and Company, 1991.

Said, Edward. *Culture and Imperialism*. New York: Vintage Books, 1993.

———. *Orientalism*. New York: Vintage, 1978.

Savage, William W. *Comic Books and America, 1945–1954*. Norman and London: University of Oklahoma Press, 1990.

Scigliano, Robert. *South Vietnam: Nation Under Stress*. Boston: Houghton Mifflin Company, 1963.

Shafer, D. Michael. *Deadly Paradigms: The Failure of U.S. Counterinsurgency Policy*. Princeton, New Jersey: Princeton University Press, 1988.

Sheehan, Neil. *A Bright Shining Lie: John Paul Vann and America in Vietnam*. New York: Vintage Books, A Division of Random House, Inc., 1989.

Shirmer, Daniel, and Stephen R. Shalom, eds. *The Philippines Reader*. Boston: South End Press, 1987.

Shulman, Holly Cowan. *The Voice of America: Propaganda and Democracy, 1941–1945*. Madison: University of Wisconsin Press, 1990.

Snyder, Alvin. *Warriors of Disinformation: American Propaganda, Soviet Lies, and the Winning of the Cold War, an Insiders Account*. New York: Arcade Publishing, 1995.

Sorensen, Thomas C. *The Word War: The Story of American Propaganda*. New York, Evanston and London: Harper & Row, 1968.

Spurr, David. *The Rhetoric of Empire: Colonial Discourse in Journalism, Travel Writing, and Imperial Administration*. Durham, North Carolina: Duke University Press, 1993.

Stevenson, William. *A Man Called Intrepid: The Secret War*. New York and London: Harcourt Brace Jovanovich, 1976.

Summers, Harry, Jr. *On Strategy: A Critical Analysis of the Vietnam War*. Novato, California: Presidio Press, 1982.

Summers, Robert E. *America's Weapons of Psychological Warfare*. New York: Arno Press, A New York Times Company, 1972.

Taylor, Keith Weller. *The Birth of Vietnam*. Berkeley and Los Angeles: The University of California Press, 1991.

Templer, Robert. *Shadows and Wind: A View of Modern Vietnam*. New York: Penguin Books, 1999.

Thayer, Thomas C. *War Without Fronts*. Boulder, Colorado: Westview Press, 1985.

The United States Information Agency: A Commemoration. Washington, D.C.: Government Printing Office, 1999.

Thomas, Nicholas. *Out of Time: History and Evolution in Anthropological Discourse*. Cambridge: Cambridge University Press, 1989.

Topmiller, Robert J. *The Lotus Unleashed: The Buddhist Peace Movement in South Vietnam, 1964–1966*. Lexington: The University of Kentucky Press, 2002.

Troy, Thomas F. *Donovan and the CIA: A History of the Establishment of the Central Intelligence Agency*. Frederick, Maryland: Aletheia Books, University Publications of America, Inc., 1981.

———. *Wild Bill and Intrepid: Donovan, Stephenson, and the Origin of The CIA*. New Haven and London: Yale University Press, 1996.

Tuch, Hans N. *Communicating with the World*. New York: St. Martin's Press, 1990.

Tuch, Hans N., and Lewis Schmidt, eds. *Ike and USIA: A Commemorative Symposium*. Washington, D.C.: The United States Information Agency Alumni Association, 1991.

Vadas, Robert E. *Cultures in Conflict: The Vietnam War*. Westport, Connecticut and London: Greenwood Press, 2002.

Vaughn, Stephen L. *Holding Fast the Inner Lines: Democracy, Nationalism and the Committee on Public Information*. Chapel Hill: University of North Carolina Press, 1980.

Wagnleitner, Reinhold. *Coca-Colonization and the Cold War: The Cultural Mission of the United States in Austria after the Second World War*. Chapel Hill and London: The University of North Carolina Press, 1994.

Weil, Martin. *A Pretty Good Club: The Founding Fathers of the U.S. Foreign Service*. New York: Norton, 1978.

Whitfield, Stephen J. *The Culture of the Cold War*. Baltimore and London: The Johns Hopkins University Press, 1991.

Whitton, John Boardman, ed. *Propaganda and the Cold War: A Princeton University Symposium*. Washington, D.C.: Public Affairs Press, 1963.

Wieck, Randolph. *Ignorance Abroad: American Educational and Cultural Foreign Policy*. Westport, Connecticut: Praeger Publishers, 1992.

Winkler, Allan M. *The Politics of Propaganda: The Office of War Information, 1942–1945*. New Haven and London: Yale University Press, 1978.

Young, Marilyn B. *The Vietnam Wars, 1945–1990*. New York: HarperPerennial: A Division of HarperCollinsPublishers, 1991.

JOURNALS

"Psychological Warfare in Korea," *Public Opinion Quarterly* 15, no. 1 (spring 1951): 65–75.

Barrett, Raymond J. "PSYOP: What Is It? What Should We Do About It?" *The PSYOP Digest* 1, no. 1 (Jan–Mar 1977): 29–48.

Bernhard, Nancy E. "Clearer than Truth: Public Affairs Television and the State Department's Domestic Information Campaigns, 1947–1952." *Diplomatic History* 21, no. 4 (fall 1997): 545–67.

Boyd, Curtis D. "Army IO is PSYOP: Influencing More With Less," *Military Review* 87, no. 3 (May–June 2007): 67–75.

Chiarelli, Peter W., and Patrick R. Michaelis. "Winning the Peace: The Requirement for Full Spectrum Operations." *Military Review* 85, no. 4 (July–August 2005): 4–17.

Currey, Cincinnatus. "Preparing for the Past." *Military Review* 69, no. 1 (January 1989): 3–13.

Frey, Marc. "Tools of Empire: Persuasion and the United States's Modernizing Mission in Southeast Asia," *Diplomatic History* 27:4 (September 2003): 543–68.

Henderson, William. "South Vietnam Finds Itself." *Foreign Affairs* 35, no. 2 (January 1957): 285–92.

Laurie, Clayton D. "Black Games, Subversion, and Dirty Tricks: The OSS Morale Operations Branch in Europe, 1943–1945." *Prologue* 25, no. 3 (fall 1993): 258–71.

———. "'The Chanting of Crusaders': Captain Heber Blankenhorn and AEF [American Expeditionary Forces] Combat Propaganda in World War I." *The Journal of Military History* 59, no. 3 (July 1995): 457–82.

Marquis, Jefferson P. "The Other Warriors: American Social Science and Nation Building in Vietnam." *Diplomatic History* 24, no. 1 (winter 2000): 79–105.

Marr, David G. "Fire in the Lake: The Vietnamese and the Americans in Vietnam." *Journal of Asian Studies* 32, no. 3 (May 1973): 564–65.

Payne, Kenneth. "The Media as an Instrument of War." *Parameters* 35, no. 1 (spring 2005): 81–93.

Petraeus, David H. "Learning Counterinsurgency: Observations From Soldiering in Iraq." *Military Review* 86, no. 1 (January–February 2006): 2–12.

Post, Jerrold M. "Psychological Operations and Counterterrorism." *Joint Forces Quarterly*, no. 37 (April 2005): 105–10.

Sanders, M. L. "Wellington House and British Propaganda During the First World War." *Historical Journal* 18, no. 1 (March 1975): 119–46.

Scanlon, J. D. "Truth: In Defense of Military Public Affairs Doctrine." *Military Review* 87, no. 3 (May–June 2007): 92–96.

Taylor, Philip M. "The Foreign Office and British Propaganda During the First World War." *Historical Journal* 23, no. 4 (December 1980): 875–98.

Trent, Stoney, and James L. Doty III. "Marketing: An Overlooked Aspect of Information Operations." *Military Review* 85, no. 3 (July–August 2005): 70–74.

Waghelstein, J. D. "Post-Vietnam Counterinsurgency Doctrine." *Military Review* 65, no. 5 (May 1985): 36–45.

Weinberg, Sydney S. "What to Tell America: The Writer's Quarrel in the Office of War Information." *Journal of American History* 55, no. 1 (June 1968): 73–89.

Index

About the Author

Robert J. Kodosky teaches courses in American history at West Chester University. He holds an MA in world history from Villanova University and a Ph.D. in United States history from Temple University. For his work on American psychological operations, Robert is a past recipient of grants from the Lyndon B. Johnson foundation, the Center for the Study of Force and Diplomacy at Temple University, and the Center for the Study of Intelligence. *Psychological Operations American Style* is his first book. Robert's current research and teaching continues to explore the intersection of diplomatic, military and cultural issues in modern American history.

www.ingramcontent.com/pod-product-compliance
Ingram Content Group UK Ltd.
Pitfield, Milton Keynes, MK11 3LW, UK
UKHW012038150926
475560UK00001B/64

9 780739 121399